COMPRESSED FOOD BARS

COMPRESSED FOOD BARS

M.T. Gillies

NOYES DATA CORPORATION
Park Ridge, New Jersey London, England
1974

Library of Congress Catalog Card Number: 74-82949
ISBN: 0-8155-0547-7
Printed in the United States

Published in the United States of America by
Noyes Data Corporation
Noyes Building, Park Ridge, New Jersey 07656

Foreword

Forward-thinking food processors wishing to enter the growing compressed food bar market will latch onto this book which comprehensively collates the results of both government and industrial research. The government research reports are the project summaries of U.S. Army Foods Laboratory studies. The industrial research material is based on patented commercial processes.

We are fortunate in the United States to be receiving direct help not only from the numerous surveys but also from active research and development programs which are being supported by the Federal Government.

This book condenses vital data that are scattered and difficult to pull together. Important processes are interpreted and explained by examples from U.S. patents. This condensed information will enable you to establish a sound background for entering the field of concentrated food manufacture.

Advanced composition and production methods developed by Noyes Data are employed to bring our new durably bound books to you in a minimum of time. Special techniques are used to close the gap between "manuscript" and "completed book." Industrial technology is progressing so rapidly that time-honored, conventional typesetting, binding and shipping methods are no longer suitable. Delays in the conventional book publishing cycle have been bypassed to provide the user with an effective and convenient means of reviewing up-to-date information in depth.

The Table of Contents is organized in such a way as to serve as a subject index and provides easy access to the information contained in this book.

Contents and Subject Index

Introduction

Dehydration of food products is well known as a procedure for preserving such products as well as a method of reducing the weight thereof. This procedure is of particular importance for military operations since it is desirable to eliminate insofar as possible all need for refrigeration equipment for preserving foods as well as to reduce the weight of foods which have to be carried by the foot soldier.

Such considerations are also important to space and land explorers, backpackers, campers, etc. However, since dehydration does not result in any appreciable reduction in volume or increases in densities of foods, it has become increasingly important to compact dehydrated foods so that they will occupy less space as well as be lighter in weight than the natural foods or foods that have been preserved by canning methods or by processing in flexible containers without dehydration.

Progress has been made which indicates that the compression of dehydrated and freeze-dried foods is a possible and even practical goal. Potential volume savings can be up to 5 to 1 with meat products and as high as 16 to 1 with some vegetables.

The term compaction has been used where the foods are compressed or otherwise made as dense as possible before dehydration. It is also used when the dry product is shaken or otherwise made to settle in the container. The term compression is used when the dry product is mechanically formed into a bar or cube with a press or pelleting machine.

Compaction is currently used with certain fruits and leafy vegetables before freeze-drying. Generally, these products are forced into a mold, frozen, and sawed into blocks for freeze-drying. Foods for astronaut feeding are carefully compacted to a uniform density to improve packaging characteristics and to standardize calorie content.

Compressed freeze-dried foods may be divided into three general classes. There are bars which are intended to be eaten as is with no rehydration, bars or blocks which are intended for use in organized dining halls or situations where equipment and time are available for rehydration, and bars intended for stress situations where the bar can be eaten as is or rehydrated in a very few minutes with either hot or cold water. All three of these classes of food bars are discussed in this book.

Recovery of Compressed Dehydrated Foods

The logistic significance of reducing the weight and bulk of military operational rations has long been recognized. In varying degrees, weight and bulk are major factors in the packaging, handling, storage and transportation of all military subsistence. These factors become critical in the design of a food packet for special missions on which a soldier must carry his entire food supply for periods up to eight days. For a great variety of foods, freeze-drying combines reliable preservation with maximum weight reduction and provides for rapid rehydration to yield products of superior acceptability. Freeze-drying, however, does not result in a consequential decrease in bulk. Previous studies have demonstrated that most freeze-dried foods, when properly plasticized, can be compressed to a volume ranging from one-third to one-twentieth of their initial volume. This compression can be achieved without a consequential amount of fragmentation and is reversible, in that on rehydration the product returns to its prefrozen size and shape with no apparent damage from the compression experience.

The material in this chapter is taken from two reports on studies made for the U.S. Army Natick Laboratories. These investigations sought to extend the knowledge of factors which influence the reversible compression of freeze-dried foods and to develop procedures to assure the reversible compression of representative dehydrated foods. The studies were under the direction of A.P. MacKenzie and B.J. Luyet of the American Foundation for Biological Research at Madison, Wisconsin, and are described in Technical Reports 70-16-FL and 72-33-FL, dated July 1969 and December 1971, respectively.

These reports describe a study undertaken: (1) to develop procedures assuring full recovery of representative dehydrated foods, after compression; (2) to describe the mechanisms operating during this recovery; and (3) to characterize irreversible changes precluding full recovery. The work was, in most respects, conducted to permit a continued comparison of different methods of preparing freeze-dried foods for compression. Special attention was devoted to the examination of the relative merits of methods based on (a) resorption from a too dry, freeze-dried state and (b) direct desorption by "limited freeze-drying." The major effort was directed to the systematic observation of the restoration of foods with reference to water activity (a_w), prior to compression, and to the measurement of the dependence of water content on a_w. The term "water activity" may be expressed in percent relative humidity. Supplementary studies were conducted to furnish evidence of the cytological and/or the physicochemical basis of the behavior during compression in selected cases. Hopefully, the value of the several methods employed has been demonstrated.

PREDRYING PROCESSES

Materials

Apples (Cortland, McIntosh, Red Delicious, Winesap and Yellow Delicious), beef (sirloin tip, graded U.S. Choice, selected for leanness), carrots (fresh, 1" to 1½" in diameter), chicken (ready-to-cook whole breasts, about 8 ounces each), cottage cheese (creamed and dry, large curd), gravy mixes (various dehydrated), mushrooms (fresh, about 1½" diameter), noodles (flat, 5.5% egg solids), pineapple (canned, unsweetened pack and fresh), potatoes (sliced dried and whole fresh), tuna (fancy, solid pack Albacore, in water), white sauce mixes (various), and additional ingredients for gravies and sauces were purchased at local supermarkets.

Red-Core Chantenay carrots and Early Frosty peas were obtained from Libby, McNeil and Libby, Inc. The carrots were delivered scrubbed, blanched, diced and inspected, and the peas were delivered shelled, blanched, size-graded and inspected.

Preparation for Freezing

Apples: Apples were peeled, cored and cut into rings ¼" thick. The rings were dipped in a 1% ascorbic acid solution, covered and drained.

Beef Stew: Beef, carrots and peas were precooked separately in boiling water for 15, 10 and 5 minutes, respectively. Potatoes were boiled in a 15% sorbitol solution for 10 minutes. The ingredients were quickly drained and mixed into the hot gravy according to the following recipe.

Ingredients	Weights
Beef, sirloin tip, lean, 1 cm cubes	500 g
Carrots, 1 cm cubes	300 g
Peas	300 g
Potatoes, dried, sliced	175 g
Gravy (beef bouillon, canned, double strength, 640 g; water, 220 g; starch, phosphorylated, 48 g; bouillon cubes (3), 12 g)	920 g

Chicken Breasts: Chicken breasts were deboned and cut into three or four strips each (being cut as much as possible parallel and perpendicular to the muscle fibers). Each piece, tightly wrapped in aluminum foil, was immersed in boiling water for 30 minutes, removed, chilled to 2°C, and cut into pieces approximately 1 cm cubed.

Cottage Cheese: Cottage cheese did not require additional preparation.

Pineapples: Pineapples were trimmed and cut into slices 1 cm thick; woody tissues were cut away. Canned, unsweetened pineapple slices (also 1 cm thick) were drained free from fluids, under cover, on several thicknesses of filter paper.

Tuna Noodle Casserole: Noodles were rehydrated and cooked for 10 minutes in a boiling 10% aqueous glycerol solution, drained and mixed with freshly heated white sauce. To this mixture were added cold, fresh mushrooms and tuna meat, cut and broken, respectively, into pieces from 1 to 3 cm across. The mixture was made according to the following recipe.

Ingredients	Weights
Mushrooms, sliced	60 g
Noodles, glycerolated	120 g
Tuna, water pack	200 g
White sauce (nonfat dry milk solids, 38 g; cornstarch, 6 g; seasoning salt, 18 g; water, 258 g)	320 g

Freezing

The various foods (single components and mixtures of same) were frozen at one or more of five different rates in accordance with requirements of particular experiments. The greatest care was always taken to prevent foods from undergoing any surface drying prior to freezing.

Slow Freezing: Each food was spread one layer thick (apple slices, beef cubes, chicken cubes, mushrooms, pineapple, potato slices, and tuna flakes) or in multilayers 1 to 2 cm thick (beef stew, cottage cheese, gravy, noodles, tuna casserole, and white sauce) on Teflon-coated aluminum trays and placed on wooden tables in a cold room maintained at -30°C or at -40° ± 1°C. In certain cases, foods were, in addition, frozen very slowly by similar exposure to still air in a room maintained at -10° ± 1°C.

Rapid Freezing: Beef, carrots, chicken, cottage cheese, noodles, pineapple and tuna were frozen by the rapid method found to be the most convenient. Powdered carbon dioxide and food were simultaneously added into an insulated container at rates arranged to provide approximately two pounds of solid carbon dioxide per pound of foodstuff. That is, the pieces of food were separated each from the other by the freezing medium. Food and refrigerant were each recovered after about 30 minutes with the aid of sieves. Beef, carrots and chicken were also frozen very rapidly by immersion, piece by piece, in liquid nitrogen.

Frozen Storage

All the foods were placed after freezing in polyethylene bags, five to ten pounds to a bag, and stored in large protective containers in a cold room at $-40^\circ \pm 2^\circ C$.

FREEZE-DRYING AND ASSOCIATED OPERATIONS

Two fundamentally different freeze-drying techniques were employed to convert foods to states suitable for compression. In some experiments the frozen materials were freeze-dried at room temperature according to a conventional procedure to moisture contents in the range 1 to 2% per 100 g dry product. These freeze-dried products were then moistened prior to compression by exposure, via the vapor phase, at 25°C, to sources of water of predetermined water activity (a_w). In other experiments frozen foods were subjected to limited freeze-drying during which the water activity was reduced to various predetermined values in the course of a single operation.

Freeze-Drying at Room Temperature

Production Experiments: Materials destined for compression/restoration experiments, cytological studies, scanning electron microscopy, and sensory evaluations were prepared in one of several apparatus of the form outlined in Figure 1.1. Foods frozen by methods previously described were placed in precooled sample chambers. The latter were then quickly attached to an apparatus, evacuated, and allowed to warm, as drying progressed, to the temperature of the surroundings (20° to 25°C). The highest possible vacuum was maintained throughout the course of each run. No attempts were made to slow or to speed the warming of the sample chamber (or of the contents) to 25°C. Thus, essentially, the materials were subjected to a mild form of commercial processing.

Rate Measurements: Indications of rates of freeze-drying at room temperature were obtained in freeze-drying equipment (shown in Figure 1.2), which incorporated a continuously recording, null-type Cahn R.H. Electro-balance with which sample weight was followed throughout freeze-drying. Heat transfer to the sample (20 g, more or less, contained in a wire mesh basket suspended in the sample chamber) was accomplished partly by radiation, partly by conduction through the vapor. Efforts were made to duplicate the conditions obtained in the production of larger quantities in the other apparatus.

Limited Freeze-Drying

The term "limited freeze-drying" was applied by Dr. A.P. MacKenzie to a process by which foods, microorganisms, and various model systems were subjected

FIGURE 1.1: APPARATUS FOR CONVENTIONAL FREEZE-DRYING

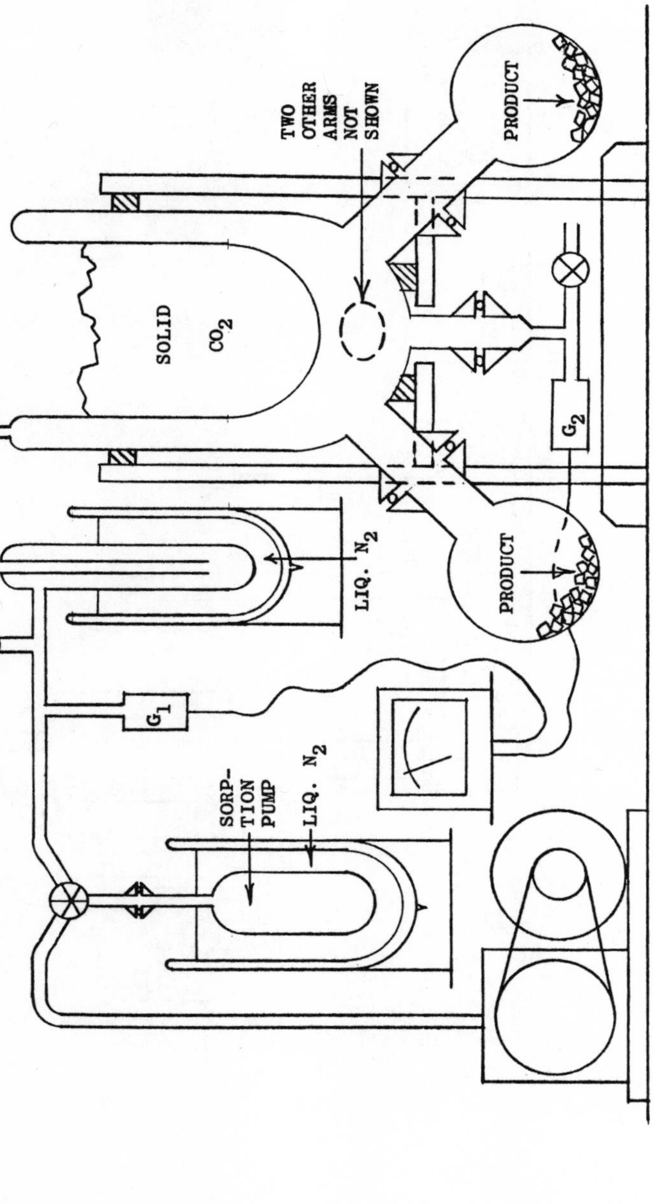

Source: Technical Report 72-33-FL, December 1971

FIGURE 1.2: APPARATUS FOR THE DETERMINATION OF FREEZE-DRYING RATES

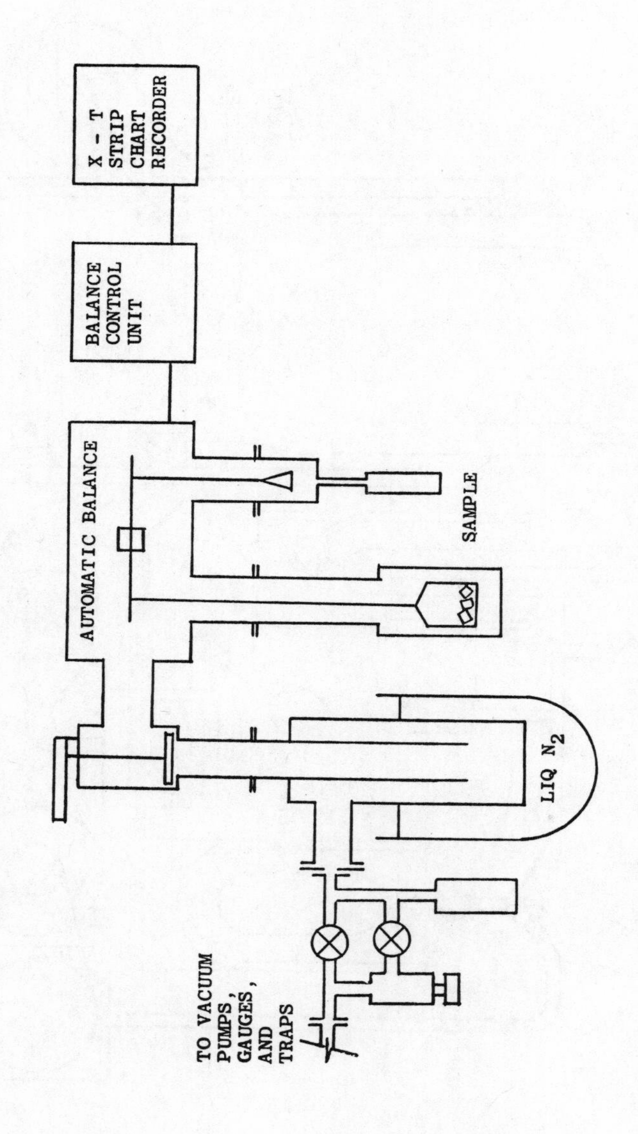

Source: Technical Report 72-33-FL, December 1971

with considerable success to differing degrees of partial dehydration at different subfreezing temperatures.

The principle of the method is evident from the block diagram reproduced in Figure 1.3. When sample chamber and condenser temperatures are separately controlled, each within narrow limits (±0.1°C each), freeze-drying proceeds by simultaneous sublimation of ice and limited desorption of unfrozen water. While the sublimation proceeds to completion, the desorption process stops when the vapor pressure of the water in the product is reduced to that of water vapor in the sample chamber in equilibrium with ice on the condenser. Greater sample chamber/condenser temperature differences result in smaller ultimate a_w's in the sample chamber; hence greater desorption from the samples.

Generally, the selection of sample chamber temperatures in the range -10° to -40°C and sample chamber/condenser temperature differences in the range of 2 to 20°C permits the production of materials containing 25 to 5 grams H_2O per 100 grams dry solids (more or less).

Production Experiments: Approximately 100 gram quantities of each of ten foods were prepared for compression/restoration studies in an apparatus of the type depicted in Figure 1.4. In each instance, foodstuffs were exposed in the sample chamber to surroundings maintained at -10° ± 0.1°C and to water-vapor pressures tending, during freeze-drying, to that of ice on a condenser maintained at -12.9° ± 0.1°C. That is, limited freeze-drying was conducted first in each case to completion at a water activity of 0.7 (±0.01). The practice of defining a_w below 0°C has been followed with reference to vapor pressures of liquid water cited by Mason (5) and reproduced in Table 1.1.

TABLE 1.1: CONDENSER TEMPERATURES PROVIDING DESIRED WATER ACTIVITIES AT SELECTED SAMPLE TEMPERATURES

	Sample Temperatures (°C)				
a_w	-10	-20	-30	-40	-50
1.00	-8.90	-17.96	-27.16	-36.50	-46.0
0.90	-10.10	-19.06	-28.20	-37.50	-46.9
0.80	-11.42	-20.30	-29.34	-38.54	-47.9
0.70	-12.89	-21.68	-30.63	-39.73	-49.0
0.60	-14.59	-23.26	-32.08	-41.09	-50.3
0.50	-16.55	-25.09	-33.80	-42.69	-51.5
0.40	-18.91	-27.31	-35.86	-44.60	-53.4
0.30	-21.91	-30.10	-38.47	-47.00	-55.7
0.20	-25.99	-33.93	-42.04	-50.4	-59.4
0.15	-28.82	-36.58	-44.53	-52.7	-61.0
0.10	-32.70	-40.20	-47.90	-55.8	-64.0
0.05	-39.03	-46.17	-53.0	-61.0	-68.7
0.025	-45.0	-51.0	-58.8	-65.9	-73.0

FIGURE 1.3: DESCRIPTION OF PARTS AND RELATIONSHIPS REQUIRED FOR LIMITED FREEZE-DRYING

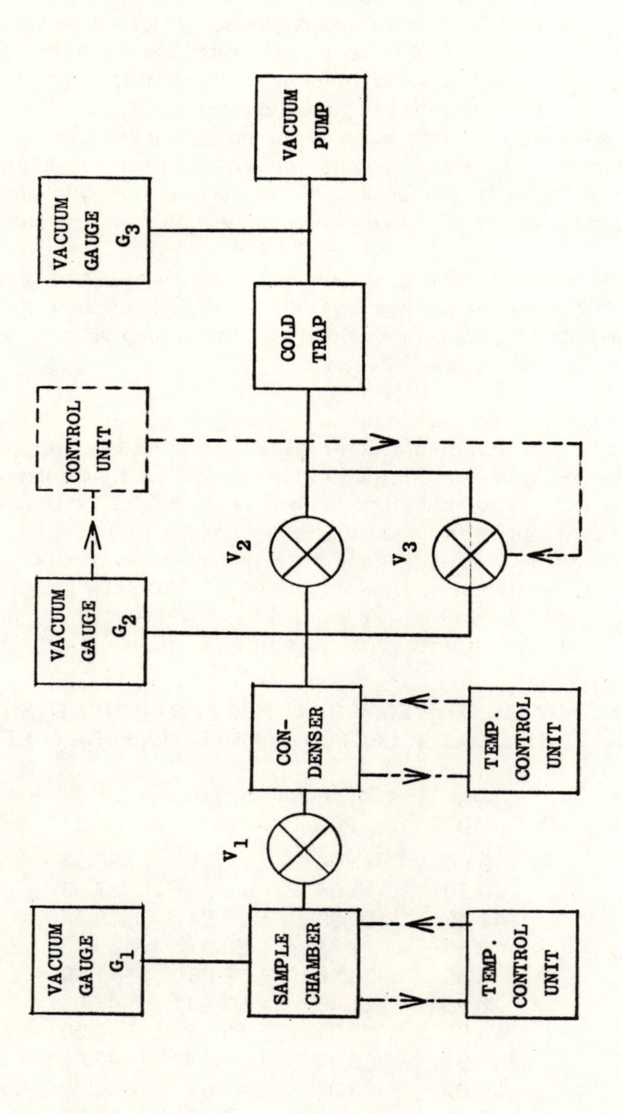

Source: Technical Report 72-33-FL, December 1971

FIGURE 1.4: APPARATUS FOR LIMITED FREEZE-DRYING

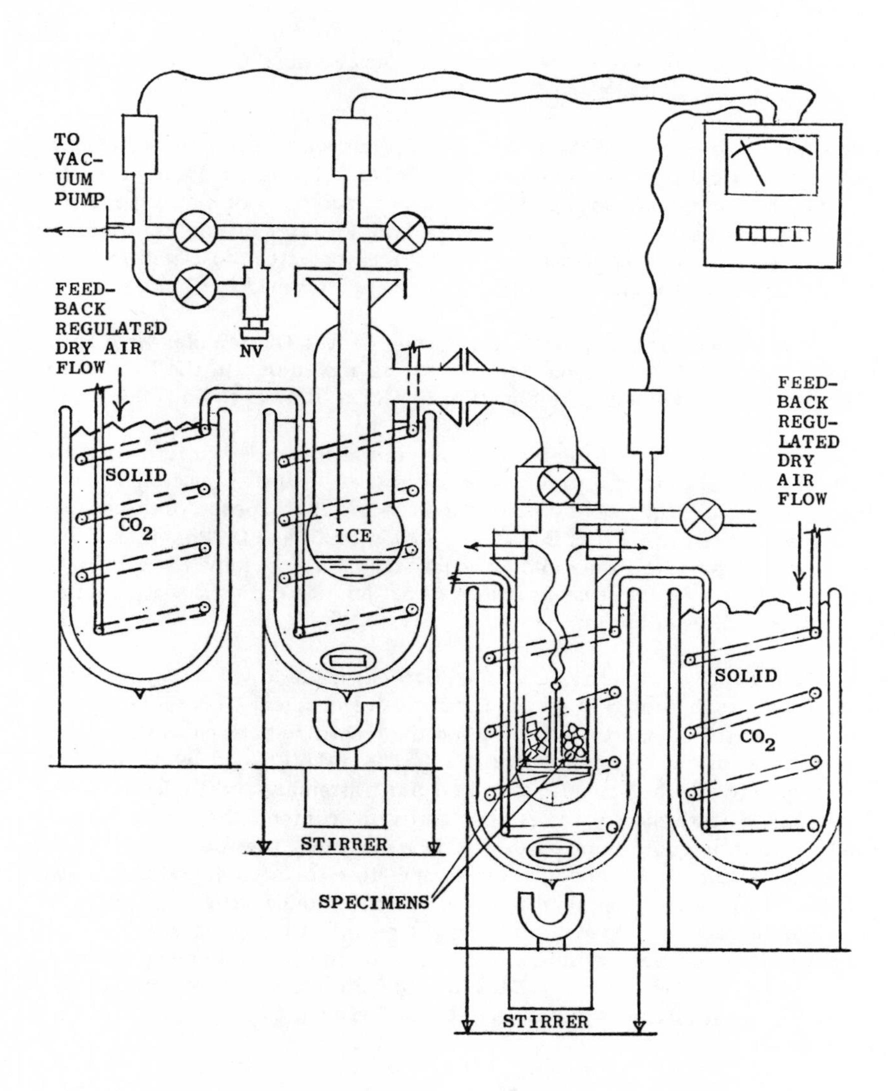

Source: Technical Report 72-33-FL, December 1971

Quantities of materials were then removed to storage. When the apparatus was reevacuated, the condenser temperature was lowered (to $-14.6^\circ \pm 0.1^\circ C$) to yield an a_W at $-10^\circ C$ of 0.6. When the ensuing further desorption was completed, additional samples were removed from the apparatus. Materials subjected to limited freeze-drying at $-10^\circ C$ and to further desorption were, in the same way, removed from the apparatus after successive stepwise equilibration to water activities of 0.5, 0.4, 0.3, 0.2 and 0.1.

Determination of Extent of Water Retention—Desorption Isotherms at -10°C: The necessary desorption experiments were made either in an apparatus of the type described in the previous section (in which materials were also prepared in greater quantities) or in an apparatus for limited freeze-drying incorporating a continuously recording balance (essentially the apparatus illustrated in Figure 1.2, modified in accordance with the requirements outlined in Figure 1.3).

In the first instance, the apparatus was opened daily. The samples were always weighed in a room at $-10^\circ C$ and returned to the apparatus. In the latter case, the weight of the sample was obtained in the form of a continuous recording.

Each product was, in any case, reduced by limited freeze-drying to a constant weight at $-10^\circ C$ appropriate to an a_W equal to 0.70. Further successive reductions to constant weight were obtained in each case by exposure, in sequence, to water activities of 0.6, 0.5, 0.4, 0.3, 0.2, 0.1, 0.05, and 0.025, all at $-10^\circ C$. Water contents based on the sample weight at 0.00 a_W (obtained at $20^\circ C$ rather than at $-10^\circ C$) were plotted as functions of a_W (and thus of relative humidity) to yield desorption isotherms.

In one case a special series of determinations was made in which a_W and temperature were each altered in a cyclic, progressive manner. This series was accomplished in the apparatus incorporating the recording balance as follows. Limited freeze-drying was completed at $-10^\circ C$ to a 0.50 a_W, at which stage the large valve between the balance/sample chamber assembly and the condenser was closed. Effectively isolated at constant water content, the sample was allowed to warm to room temperature. The water vapor pressure corresponding to the equilibrium established at room temperature was noted, the sample was cooled to $-10^\circ C$, and the equilibrium pressure noted once again. The valve to the condenser was then opened, insuring the return of the sample to 0.50 a_W. Changes in sample weight arising in the course of the four-step cycle were noted. Desorption was then extended, at $-10^\circ C$, to a 0.30 a_W, at which stage the sample was exposed to the same isolation/warming/cooling/reconnection procedure.

Desorption was subsequently extended at $-10^\circ C$ to water activities of 0.30, 0.09 and 0.05, at each of which the sample was exposed to the isolation/warming/cooling/reconnection procedure. Changes in water vapor pressure and sample weight were in each case noted.

Humidification

Foods freeze-dried in the usual way to low water contents were moistened prior to compression by exposure to water vapor as follows.

Resorption Isotherms: Quantities of each of twelve test materials freeze-dried at room temperature were taken from storage over dry Linde Molecular Sieve, weighed and distributed between eight vacuum desiccators, each partly filled with aqueous sulfuric acid. The weights of the twelve materials thus exposed, in vacuo, to solutions having water activities of 0.10, 0.20, 0.30, 0.40, 0.50, 0.60, 0.70 and 0.80 were determined periodically, also the times beyond which these weights did not change further. Later the samples were exposed a second time to dry Linde Molecular Sieve to permit the redetermination of the dry weight. Water resorbed per unit weight of dry product at 25°C was plotted as a function of water activity in each case.

Preparation for Compressibility Studies and for Sensory Evaluation: Larger quantities of each material were similarly prepared for studies on the effects of different degrees of resorption on compression and on subsequent recovery. Resorption was allowed to proceed for times equal to those taken by samples to reach constant weight in previous tests. Quantities of certain foods destined for taste panel studies were prepared in numbers of large desiccators in much the same way with the following differences: (a) sulfuric acid solutions generating precisely the required water activities were poured into the desiccators, (b) weights of water equal to those to be taken up by the foods (the values were obtained from the dry weights of the foods and the resorption isotherms) were added slowly to the sulfuric acid solutions, and (c) the sulfuric acid was stirred slowly and continuously by means of magnets after the foods were sealed in the desiccators.

Rate Measurements: The freeze-drying apparatus incorporating the continuously recording balance to measure humidification velocities was used. Atmospheres of required water activity were obtained by suitable choice of condenser temperature. Resorption was initiated when the valve between the sample chamber containing the fully dried specimen and the condenser (now acting as source of water) was opened.

Compression

Physical Compression/Restoration Studies: Moistened freeze-dried foods were compressed in small quantities (several grams on a dry weight basis) between aluminum plates mounted in a hydraulic press. Pressures employed were calculated on the basis of observed pressure readings and measurements of product area after compression. Pressures were maintained at certain constant values for periods, generally, of one minute.

Sensory Evaluation: Compression of larger quantities was accomplished 250 ml at a time in an aluminum cylinder having smooth-faced aluminum discs, closely fitting the cylinder walls, for sliding ends (Figure 1.5). The cylinder was filled with the freeze-dried food having the desired moisture content, closed and inserted in the press. Each material was subjected to 500 psi of sample area for one minute, after which the pressure was quickly released.

FIGURE 1.5: COMPRESSION ASSEMBLY

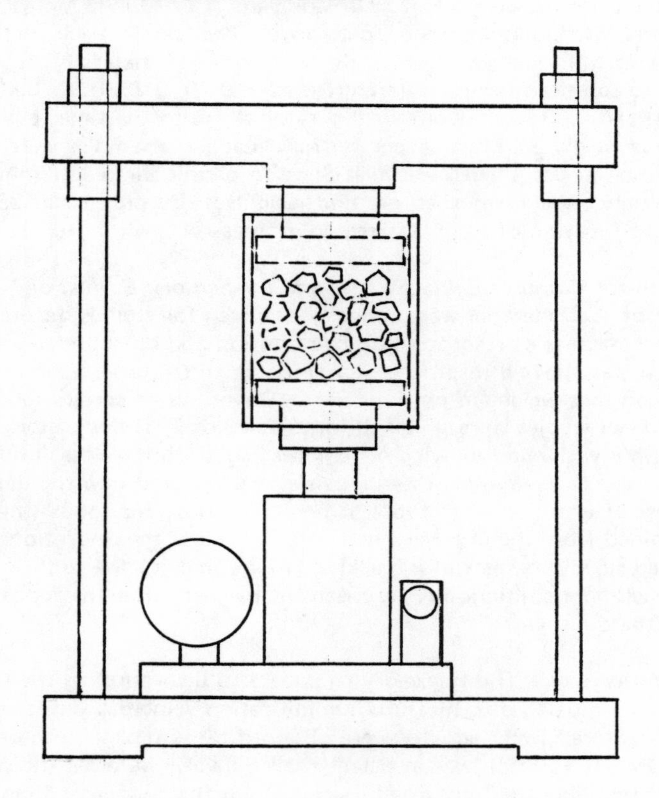

Source: Technical Report 72-33-FL, December 1971

Final Drying and Storage

Drying and Storage: Compressed materials were dried in vacuum desiccators over dry Linde Molecular Sieve, at 25°C. Bulk densities before and after final drying were determined from measurements of sample area, thickness and weight. Final storage was effected in vacuum, over dry Linde Molecular Sieve, at 2°C.

Rate Measurements: Final drying rates were determined immediately when resorption was completed. Resorbed specimens were removed from the apparatus containing the recording balance, subjected to 500 psi for one minute, and returned to the suspended basket. The condenser temperature was lowered to -196°C, the sample being kept meanwhile at 25°C. Wide-bore valves were open throughout the system during final drying, thus insuring the completion of the process under high vacuum.

Recovery

Pieces of various foods at different water contents were subjected to various pressures for different periods of time, to final drying in some cases, and to rehydration in water. Each sample's behavior was noted with reference to orientation, pressure, duration of compression, and to the method used to permit restoration (flotation vs forced immersion; hot water vs cold water). Foods were examined one at a time and, afterwards, in appropriate admixture.

Cytological Studies

To determine the combined effect of the compression and restoration procedures on the microscopic structure, certain foods were subjected to conventional cytological study. Foods were cooked (where necessary), frozen, freeze-dried, humidified to various predetermined extents, compressed, subjected to final drying and, lastly, to rehydration (in hot or cold water, whichever resulted in better recovery). The rehydrated materials were fixed in formalin-acetic acid-ethyl alcohol (FAA), dehydrated in graded mixtures of water, ethanol and n-butanol, and infiltrated with paraffin, MP 56° to 58°C, at 60°C.

10, 15 and 20 micron sections were cut in longitudinal and transverse directions with steel blades on a rotary microtome at 25°C. Ribbons consisting of serial sections were attached to glass slides, stained, examined and photographed, principally at a magnification of 100X. Generally the procedures described by Jensen (1) and Sass (6) were used.

Control experiments were conducted in each case on fresh and/or cooked materials, and on fresh and/or cooked materials subjected also to freezing, freeze-drying and rehydration.

Studies on the Effects of Solvent Extraction

Experiments were undertaken to determine the behavior of the various insoluble structural components of different foods, during compression, in conditions in which certain components were absent. Solvent extraction was used to effect the removal of lipids, after freeze-drying, from cooked beef and cooked chicken. Sugars and other water-soluble substances were extracted from freeze-dried, fresh apple and freeze-dried, canned pineapple. The methods were as follows.

Cooked beef and chicken were cut into 1 cm cubes, subjected to slow freezing (in air at -40°C), freeze-drying at room temperature and to 48 hour Soxhlet extraction with chloroform/methanol mixtures (2:1, v/v). Solvent extracted materials were stored in vacuo in desiccators containing Type 13X Linde Molecular Sieve in which conditions any solvent molecules remaining in the food were transferred to the sieve.

Apple and pineapple were frozen in air at -40°C in the form of thin (0.5 to 1.0 cm) slices, freeze-dried at room temperature and subjected to repeated extraction with stirred aqueous ethanol (40:60, v/v) at 15°C. The water-alcohol mixture was changed four times at daily intervals and replaced after the last change with distilled water. After a further interval, the tissues were removed, drained, frozen in liquid nitrogen, and freeze-dried a second time.

The beef, chicken, apple and pineapple were then placed in desiccators containing aqueous sulfuric acid solutions adjusted to yield, upon equilibration, a_w's of 0.20, 0.30, 0.40, 0.50, 0.60, 0.70 and 0.80 at 25°C. When the desiccators were opened, the samples were subjected to pressures of about 500 psi for periods of one minute and examined for tendencies to recover original shapes, sizes and textures in water.

RESULTS

Preliminary Research

Small scale studies were conducted to determine best variety, best gravy, best sauce and best recipe, where necessary. Best choice was judged on the basis of response to freezing, freeze-drying, moistening, compression, drying and reconstitution. Physical compression/recovery tests and informal four-man taste panel studies determined:

- (a) that Winesap apples offered performance superior to that of Cortlandt, Delicious (Red or Yellow), or McIntosh;
- (b) that creamed cottage cheese was preferable to the dry curd product and large curd better than small curd;
- (c) that the best gravy contained the least sodium chloride and the most protein and soluble starch consistent with flavor requirements and absence of a whitening effect;
- (d) that the most suitable potato was obtained by the rehydration of dried potato slices in aqueous sorbitol solution, prior to freezing;
- (e) that the best white sauce included the least possible sodium chloride, the least possible cornstarch and the most nonfat dry milk solids consistent with the required stiffness and opacity;
- (f) that the most acceptable noodles resulted from the exposure, prior to freezing, of egg noodles to aqueous glycerol; and
- (g) that the best beef stew and tuna noodle casserole were obtained from combinations of ingredients in proportions outlined earlier in this report.

Special efforts were made to formulate gravies and white sauces having high collapse temperatures, these temperatures being determined according to criteria developed by MacKenzie (3)(4). Numerous preparations were, that is, examined in the freeze-drying microscope to determine whether or not there existed in every case a freeze-drying temperature below which freeze-drying progressed with complete retention of the solute distribution characteristic of the frozen state.

In each case the gravies and sauces underwent freeze-drying by one of two mechanisms depending on the temperature; that is, the dehydration progressed with the retention of the solute matrix structure below a certain freeze-drying temperature and with collapse above it. A gravy satisfactory from other points of view could not be obtained with a collapse temperature higher than -32°C, nor was it possible, on a similar basis, to formulate a white sauce that retained its matrix when freeze-dried above -27°C.

It was, however, observed that different gravies, freeze-dried below their respective collapse temperatures of -32°C or thereabouts, responded differently to exposure to 25°C and water activities in the range 0.4 and 0.8. High salt or high sodium glutamate mixtures collapsed on moistening to 0.5 a_w or more; starch-boullion gravies collapsed at 0.6 a_w and higher, and starch-nonfat dry milk solids-beef broth mixtures survived exposures to 0.7 a_w without collapse.

Compression/Restoration Experiments: Behavior with Respect to Water Activity

The results of the laboratory tests combining physical and subjective evaluation by two assistants are presented in Tables 1.2 and 1.3. Behavior is listed according to food, method of preparation, water activity and, in some cases, freezing rate. A_w's were measured at the temperatures at which the respective sorption equilibria were established (25°C for resorption, -10°C for desorption). Table 1.2 describes the reactions of foods frozen slowly prior to freeze-drying at room temperature; Table 1.3 describes the behavior of foods prepared by slow and rapid freezing and by limited freeze-drying.

Good restoration was observed in cooked white chicken muscle after slow freezing but not after rapid freezing. When good recovery was observed it was found in each case at only one water activity, i.e., recovery was strongly a_w dependent.

An acceptable restoration was likewise obtained with specially processed sliced potatoes at one a_w. Performance was, however, very strongly dependent on a_w (Table 1.2 only). Several varieties of fresh mature potatoes lost much of their respective textures after freeze-drying and rehydration and further proved to be quite unsuitable for compression.

Beef stew, judged as a composite dish, recovered an adequate overall quality somewhat less dependent on a_w prior to compression than performance of individual components. Limited restoration in one component (beef, carrots, peas, potatoes) was perhaps less noticeable from a subjective point of view where

TABLE 1.2: EFFECTS OF WATER ON FREEZE-DRIED FOODS COMPRESSED AFTER REMOISTENING TO VARIOUS WATER ACTIVITIES

(Freeze-dried materials were exposed to atmospheres of these water activities at 25°C prior to compression at 25°C.)

a_w	Chicken (Cooked)	Potatoes (Cooked)	Beef Stew
0.80	Compresses easily; fails to restore at all.	–	–
0.70	Compresses easily; fair restoration.	Compress readily but fail in part to resorb and detach.	Gravy collapsed on other components prior to compression; minimal restoration; gravy hard to wet.
0.60	Compresses and restores well.	Compress and restore fairly well.	Compresses well; all components restore; carrots and peas best; meat slightly tough; potatoes soft.
0.50	Compresses but disintegrates in part upon rehydration.	Slices shatter upon compression; some pieces restore.	Compresses well; components separate well on rehydration; meat tender; peas and potatoes tend to disintegrate.
0.40	Crumbles upon compression.	Fragment upon compression; no restoration.	Peas, potatoes crumble; mixture hard to compress; poor recovery.
0.30	Fragments upon compression.	–	–
0.20	Powders upon compression.	–	–

a_w	Mushrooms (Cooked)	Noodles (Cooked)	Tuna (Cooked)	Tuna-Noodle Casserole
0.80	Essentially no recovery.	Compress well; fair recovery; tendency to agglutinate.	Compresses readily; good recovery.	Good compression of all components; good separation and recovery; noodles tend not to disperse.
0.70	Very limited recovery.	Compress well; restore well save for tendency to clump.	Tends to crack upon compression; good recovery.	Good compression of all components; good separation and recovery; noodles tend not to disperse.

TABLE 1.2: (continued)

a_w	Mushrooms (Cooked)	Noodles (Cooked)	Tuna (Cooked)	Tuna-Noodle Casserole
0.60	Limited recovery.	Tendency to break upon compression; fair restoration.	Tends to break up on compression; pieces recover well, however.	Definite tendency to crush with loss of identity; less tendency for noodles to clump on rehydration; good recovery.
0.50	Fair recovery (not complete).	Considerable fragmentation; little recovery.	Fragments during compression; pieces recover well, however.	–
0.40	Limited recovery.	Crumbles upon compression.	Crumbles upon compression; recovers with fair texture.	–
0.30	Damage upon compression; very limited recovery.	–	–	–
0.20	Fragments upon compression.	–	–	–

a_w	Cottage Cheese	Pineapple	Apple
0.80	Recovers slowly to fair texture.	–	–
0.70	Recovers to good texture.	–	–
0.60	Recovers to fair texture.	–	–
0.50	Crumbles upon compression; restores to soft texture.	Collapses upon resorption; flows upon compression; disintegrates on rehydration.	Fails to recover in cold water; restores in hot water.
0.40	Fragments on compression; rehydrates to a paste.	Disintegrates on rehydration.	Fair restoration; best in warm water.
0.35	–	Fragments on rehydration.	–
0.30	Powders on compression.	Pieces intact on rehydration; texture soft.	Recovery excellent in hot or cold water.
0.25	–	Firm texture on rehydration.	–
0.20	–	Fair texture on rehydration; pieces tend not to separate.	Fair to good restoration; best in hot water.
0.15	–	Some crumbling on compression; fair texture in hot water.	Compresses but disintegrates on rehydration.

TABLE 1.2: (continued)

a_w	Cottage Cheese	Pineapple	Apple
0.10	–	Crumbles upon compression.	Crumbles upon compression.
0.05	–	Shatters upon compression.	Powders upon compression.

TABLE 1.3: EFFECTS OF WATER ON FOODS COMPRESSED AFTER LIMITED FREEZE-DRYING TO VARIOUS WATER ACTIVITIES*

a_w	Tuna (Cooked, Slowly Frozen)	Tuna (Cooked, Rapidly Frozen)	Noodles (Cooked, Both Freezing Rates)
0.65	Compresses readily; excellent restoration; rapid water uptake.	Compresses readily; excellent restoration; rapid water uptake.	Flow on compression; poor restoration; pieces tend not to separate.
0.50	Compresses well; good restoration.	Compresses well; fair restoration.	Fair restoration.
0.40	Compresses well; good restoration.	Compresses well; fair restoration.	Fair restoration.
0.30	Cracks form on compression; fair restoration.	Cracks form on compression; fair restoration.	Good restoration.
0.20	Some crumbling on compression; some restoration.	Some crumbling on compression; some restoration.	Poor restoration.
0.10	Fragments on compression; poor restoration.	Fragments on compression; poor restoration.	Fragments on compression; poor restoration.

a_w	Chicken (Cooked, Slowly Frozen)	Chicken (Cooked, Rapidly Frozen)	Cottage Cheese (Slowly Frozen)	Cottage Cheese (Rapidly Frozen)
0.65	Poor to fair restoration; pieces stick together.	Poor restoration; pieces stick together.	–	–
0.50	Poor to fair restoration.	Poor restoration.	–	–
0.40	Good restoration.	Poor restoration.	Ready compression; poor restoration.	Ready compression; poor restoration.
0.30	Fair restoration.	Poor restoration.	Ready compression; poor restoration.	Satisfactory compression; fair recovery.
0.20	Fair restoration.	Fair restoration.	Fair restoration.	Good restoration.
0.10	Tissues crack on compression; fair restoration nonetheless.	Tissues crack on compression; fair restoration nonetheless.	Powders on compression; no recovery.	Powders on compression; no recovery.

*Water activities were determined at -10°C, that being the temperature at which desorption was conducted; compression was conducted at 25°C after warming. The a_w below 0°C is defined with reference to vapor pressures of liquid water cited by Mason (5).

other components recovered well. Most persistent among the shortcomings was a tendency for potato to disintegrate in the gravy phase, the latter acquiring the consistency of a heavy paste. In summary, the several components of the beef stew were observed to restore as well (or, as poorly) in the presence of each other and the gravy solubles as they did in separate tests.

Mushrooms, subjected to 500 psi, were not observed to restore very well in any instance. Good recovery was, however, obtained where compression was limited to pressures in the range of 50 to 100 psi at 0.5 or 0.4 a_W.

Good behavior was observed in glycerolated noodles to span a rather broad range of water activities. Heavier glycerolation (20% vs 10%) resulted in a compression/recovery performance even less dependent on a_W. Taste of glycerol, however, was determined to be objectionable in the latter case. When less heavily glycerolated noodles were examined, piece-to-piece adhesion appeared to be more of a problem than restoration of individual pieces.

Slowly frozen tuna restored very well indeed. Good recovery was not critically dependent on water activity prior to compression. Tuna judged excellent in texture was obtained from tissues freeze-dried at room temperature and exposed to a_W's of 0.8, 0.7 and 0.6 prior to compression. Tissues subjected to slow freezing and to limited freeze-drying to water activities (at -10°C) of 0.6, 0.5 and 0.4 likewise restored very well. Rapid freezing prior to limited freeze-drying appeared to reduce the chances of a good recovery (cf chicken).

Tuna noodle casserole restored very well indeed. Water activities of 0.7 and 0.6 were equally effective in treatment prior to compression. Each component appeared to rehydrate and recover as well in the presence of the others as in their absence.

Good behavior was noted in creamed cottage cheese to center about an adjustment to 0.7 a_W at 25°C and to 0.3 or 0.2 at -10°C. Rapid freezing prior to limited freeze-drying was, apparently, preferable. In a series of experiments not shown in Table 1.1, dry curd cottage cheese was found to exhibit best behavior following resorption to 0.7 a_W.

Pineapple did not recover well in cold water. Rehydration with hot water, however, resulted in very good restoration (apparently without any loss in flavor). Low water activities were required to achieve best behavior.

Apples restored very well in cold water following exposures to 0.3 a_W prior to compression. Good restoration, less critically dependent on a_W, was obtained when rehydration was performed with hot water. Exposure to hot water, however, resulted in some softening.

Organoleptic Evaluation

Six foods, prepared in sufficient quantities, were rehydrated for submission to twelve man taste panel sessions, who graded them according to the following scale: 7 = like very much; 6 = like moderately; 5 = like slightly; 4 = neither like nor dislike; 3 = dislike slightly; 2 = dislike moderately; 1 = dislike very much. Some foods recovered readily and were examined by the panel. In other cases, food slabs 1 to 2 cm thick proved so resistant to rehydration they were not submitted to the panel. Where foods were not submitted to the panel, they were subjected to special examination by the kitchen staff associated with the taste panel operation.

Apples: The difficulties encountered in the rehydration of discs 1.5 cm thick precluded taste panel tests. Freeze-dried controls, however, rehydrated very readily to a most acceptable texture, slowly frozen material in 3 minutes, rapidly frozen in 2 minutes. Compressed materials exposed to cold or to very hot water underwent a total disintegration at the outer surface before the innermost areas were completely rehydrated. Intermediate zones of considerable proportion could, at the same time, be said to consist of acceptable material. The problem seemed to reside in a failure to effect a uniformly good recovery.

Beef Stew: Slowly frozen, freeze-dried control and compressed materials were rehydrated with boiling water (3 g per g dry weight) for such times as proved necessary, i.e., for 10 to 15 minutes and for 30 minutes, respectively. Some parts of some discs especially resistant to rehydration were forcefully separated. The adhesion of the potatoes, during rehydration, to the other components was particularly noticeable. The products were judged as follows.

TABLE 1.4

Mean Scores:	
Control, freeze-dried, not compressed	4.5
Resorbed to 0.50 a_w, 25°C, compressed	4.4
Resorbed to 0.60 a_w, 25°C, compressed	4.4
Resorbed to 0.70 a_w, 25°C, compressed	4.5

Chicken: Freeze-dried controls and compressed materials were restored with a slight excess of boiling water and kept in a hot water bath for 10 to 20 minutes, sufficient in each case for complete rehydration. Materials derived from slowly frozen chicken were assessed by the panel as follows.

TABLE 1.5

Control, freeze-dried, not compressed	3.5
Resorbed to 0.50 a_w, 25°C, compressed	5.4
Resorbed to 0.60 a_w, 25°C, compressed	5.2
Resorbed to 0.70 a_w, 25°C, compressed	4.9

Materials derived from rapidly frozen chicken were evaluated in a subsequent session as follows.

TABLE 1.6

Control, freeze-dried, not compressed	5.7
Resorbed to 0.50 a_w, 25°C, compressed	5.0
Resorbed to 0.60 a_w, 25°C, compressed	4.3
Resorbed to 0.70 a_w, 25°C, compressed	4.6

The slowly and the rapidly frozen materials cannot be compared, having been examined by the panel on different occasions. The definite impression was, however, gained that the slowly frozen product restored to a better texture. The rapidly frozen materials were, moreover, less readily hydrated.

Cottage Cheese: Since the compressed materials prepared in the form of discs 1 to 1.6 cm thick did not rehydrate completely in a one hour period, they were not submitted to the taste panel. Certain observations were, however, recorded during the various attempts at restoration. None of the compressed materials was totally resistant to rehydration at 25°C. None of the rehydrated materials, moreover, lost the texture regained on restoration. That is, texture, once recovered, was maintained at 25°C for periods of from one to three hours. Water entered the discs more rapidly the lower the water activity to which the freeze-dried material was resorbed prior to compression. Slow freezing prior to freeze-drying was more effective in promoting rehydration than was rapid freezing.

TABLE 1.7: EXTENT OF REHYDRATION IN THIRTY MINUTES AT 25°C*

Water Activity During Resorption Prior to Compression	Slowly Frozen Prior to Freeze-Drying	Rapidly Frozen Prior to Freeze-Drying
0.60	20 to 30%	40%
0.70	30%	50%
0.80	40 to 50%	60 to 70%

*Thickness of dry core expressed as percentage of original thickness of compressed material.

Pineapple: Slowly frozen materials were rehydrated in 100°C water without difficulty, drained, cooled and submitted to the panel. Rapidly frozen, freeze-dried, compressed products were observed not to rehydrate completely in any instance. The latter samples were therefore submitted only to brief examination. Slowly frozen, compressed, restored materials were scored by the panel as follows.

TABLE 1.8

Control, freeze dried, not compressed	4.0
Resorbed to 0.20 a_w, 25°C, compressed	2.8
Resorbed to 0.25 a_w, 25°C, compressed	4.5
Resorbed to 0.30 a_w, 25°C, compressed	4.1

The control rehydrated in seven minutes, those resorbed to 0.2, 0.25 and 0.30 a_w prior to compression in 25, 15 and 11 minutes, respectively. The panel commented favorably on the flavor retained in each instance. Rapidly frozen material resorbed to 0.10 or to 0.20 a_w prior to compression was only partly rehydrated in 30 minutes at 100°C. Very dense, tough, somewhat flexible centers persisted in otherwise unacceptably soft tissues. Material resorbed to 0.30 a_w prior to compression rehydrated in part to yield expanded discs, the outer portions of which exhibited very acceptable texture. Dense, tough, innermost zones were, however, also detected. Rapidly frozen control samples, not compressed, rehydrated to an acceptable texture in 10 minutes (cf 7 minutes for the slowly frozen controls).

Tuna-Noodle Casserole: Compressed and control materials were rehydrated with boiling water and maintained thereafter at the boiling point for periods of 30 minutes. Portions served to the panel were rated as follows.

TABLE 1.9

Control, freeze-dried, not compressed	4.9
Resorbed to 0.50 a_w, 25°C, compressed	4.6
Resorbed to 0.60 a_w, 25°C, compressed	4.3
Resorbed to 0.70 a_w, 25°C, compressed	3.6

It was observed that the compressed materials all offered some resistance to rehydration. Noodles appeared not to detach from each other. Samples rehydrated to 0.50 a_w prior to compression resorbed with the least difficulty.

Water Sorption Studies

Resorption Isotherms Obtained Subsequent to Conventional Freeze-Drying: Resorption data were collected in the form of curves denoting the binding of water by freeze-dried materials exposed to atmospheres of precisely controlled water activity. A representative plot may be seen in Figure 1.6, indicated by the letter R. All the resorption plots were obtained from foods frozen slowly prior to freeze-drying, except where indicated.

Desorption Isotherms: A representative plot is shown in Figure 1.6, indicated by the letter D. Data for these isotherms represent the extents to which the freeze-dried materials continued to bind water after limited freeze-drying to the various water activities indicated. Only in one case did the freezing rate appear to affect the course of the desorption isotherm—only in pineapple was the resorption isotherm observed (quite unexpectedly) to cross the desorption isotherm.

Special Study of Construction of a Virtual Desorption Isotherm: The extent to which the activity of the water retained during desorption increases with increased temperature is illustrated in Figure 1.7. These effects from changing temperature are indicated by the unbroken line; the direction of the changes

FIGURE 1.6: SORPTION ISOTHERMS FOR FREEZE-DRIED, CANNED, WATER PACK PINEAPPLE

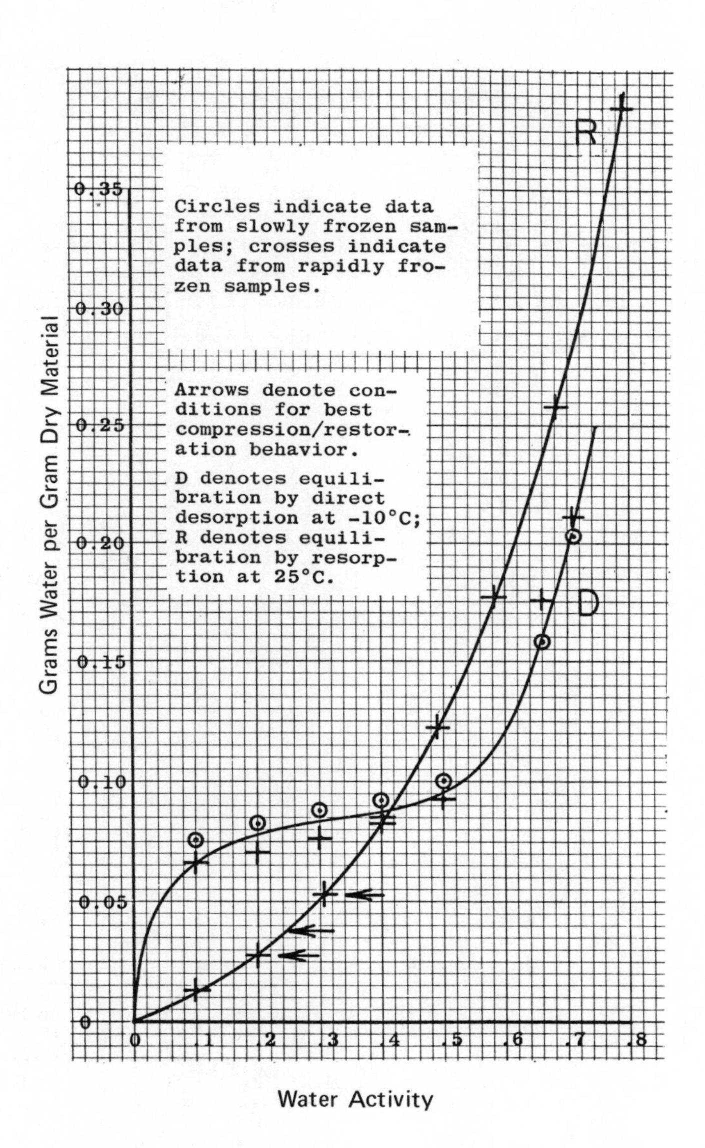

Source: Technical Report 72-33-FL, December 1971

FIGURE 1.7: VIRTUAL SORPTION ISOTHERM OBTAINED FROM FREEZE-DRIED BEEF, SLOWLY FROZEN

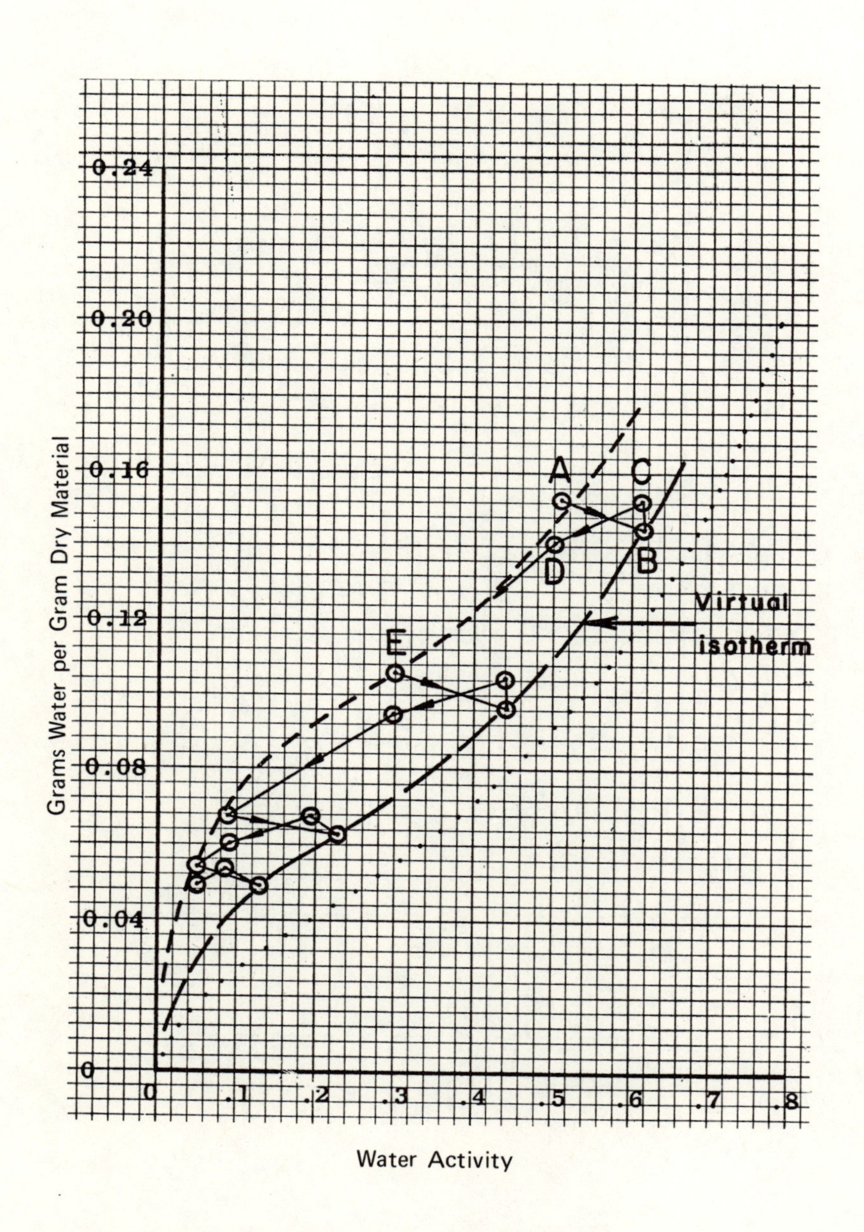

Source: Technical Report 72-33-FL, December 1971

in water activity, weight and temperature are indicated by the arrows. The virtual desorption isotherm obtained from freeze-dried beef at room temperature (23°C) was obtained by joining those points representing water activities assumed by the sample when warmed at various water contents from -10° to 23°C. This latter isotherm is represented by a line of long dashes. A line of short dashes indicates the course of direct desorption conducted entirely at -10°C. Similarly, a dotted line indicates the course of resorption at 25°C.

The behavior of the sample during the repeated desorption/warming/cooling sequences was found to be much closer to the direct desorption at -10°C than to the resorption at 25°C. One notes also that water released by the specimen on warming (from A to B for example) is regained readily when the specimen is cooled again to -10°C (B to C) but that it is released a second time (C to D) when the sample is caused once more to adjust to the water activity maintained previously at -10°C. Desorption from D to E clearly follows very nearly the course described by direct desorption at -10°C in experiments in which the sample was not subjected to intermittent warming.

Rate Determinations

Freeze-Drying at Room Temperature: Measurements of the rates at which apple, beef stew, chicken, cottage cheese, pineapple and tuna-noodle casserole freeze-dry at room temperature were made by continuous recordings obtained during each experiment. It was noted that rapidly frozen chicken freeze-dried to a lower solids content than did the slowly frozen product, and also that cottage cheese freeze-dried to different solids contents depending on the preparation. The curve for freeze-drying of cottage cheese is shown in Figure 1.8.

Tuna-noodle casserole freeze-dries as rapidly as do chicken and cottage cheese despite the presence of components rich in starch. Apples, pineapple and beef stew, by contrast, freeze-dry much less rapidly.

Humidification Velocities: The times taken by the various foods freeze-dried at room temperature to resorb water from the vapor phase were obtained during exposure of each freeze-dried food to that a_W required for best compression and restoration behavior. Figure 1.9 shows the uptake of water from the vapor by freeze-dried beef stew at 25°C. Times taken to regain nine-tenths of the water eventually resorbed varied from 3, 5.5, 6 and 7 hours for chicken rapidly frozen, slowly frozen, beef stew and cottage cheese, respectively to 15, 16 and 20 hours for tuna-noodle casserole, apple and pineapple.

Final Drying Velocities After Compression: Figure 1.9 shows the drying velocity of beef stew after compression. Chicken desorbed very rapidly; apple, beef stew and tuna casserole somewhat less rapidly, appearing reluctant to release the last 0.5, 2.5 and 1.0 gram water per 100 grams dry product, respectively. Pineapple was observed to undergo final drying from the compressed state with extreme reluctance.

FIGURE 1.8: WEIGHT/TIME CURVES DESCRIBING THE FREEZE-DRYING OF CREAMED AND DRY CURD COTTAGE CHEESE*

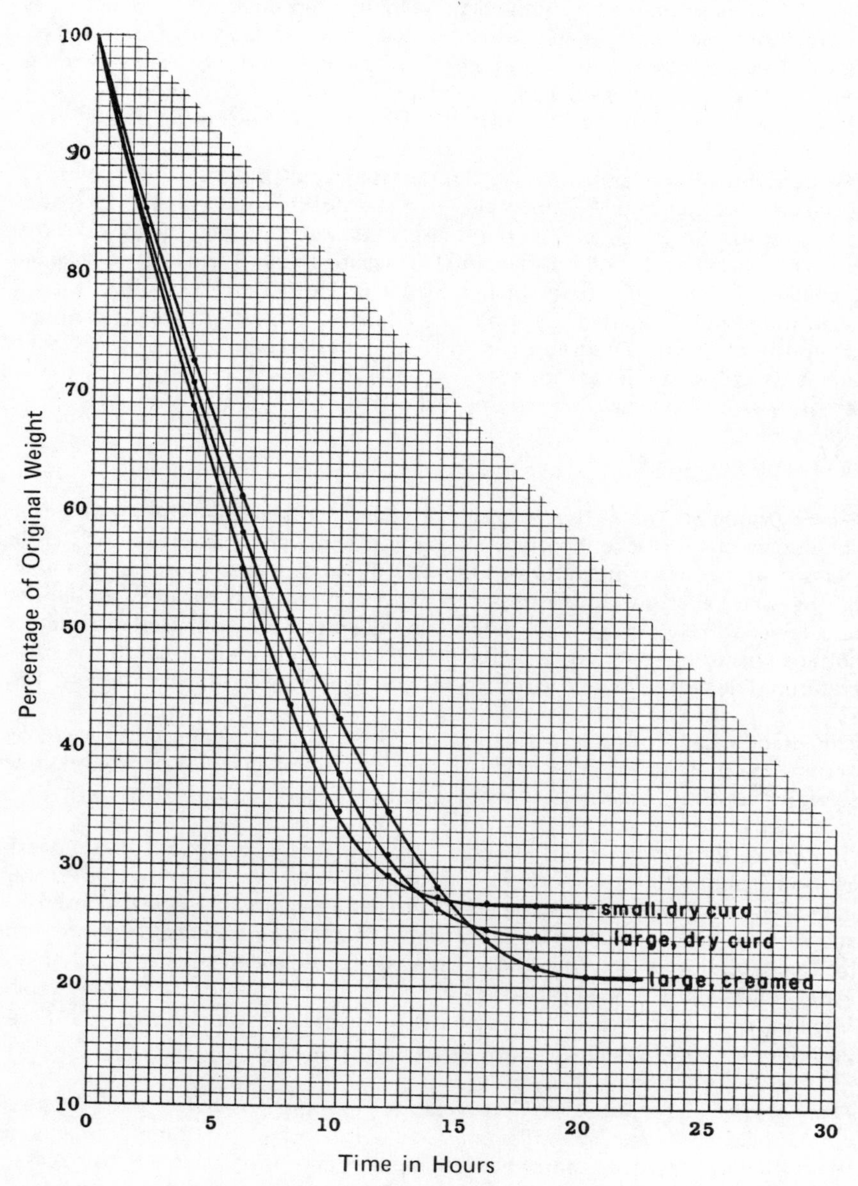

*Each slowly frozen to 0 a_w at 25°C as indicated.

Source: Technical Report 72-33-FL, December 1971

FIGURE 1.9: WEIGHT/TIME CURVES DESCRIBING THE RESORPTION AND FINAL DRYING OF FREEZE-DRIED BEEF STEW, SLOWLY FROZEN

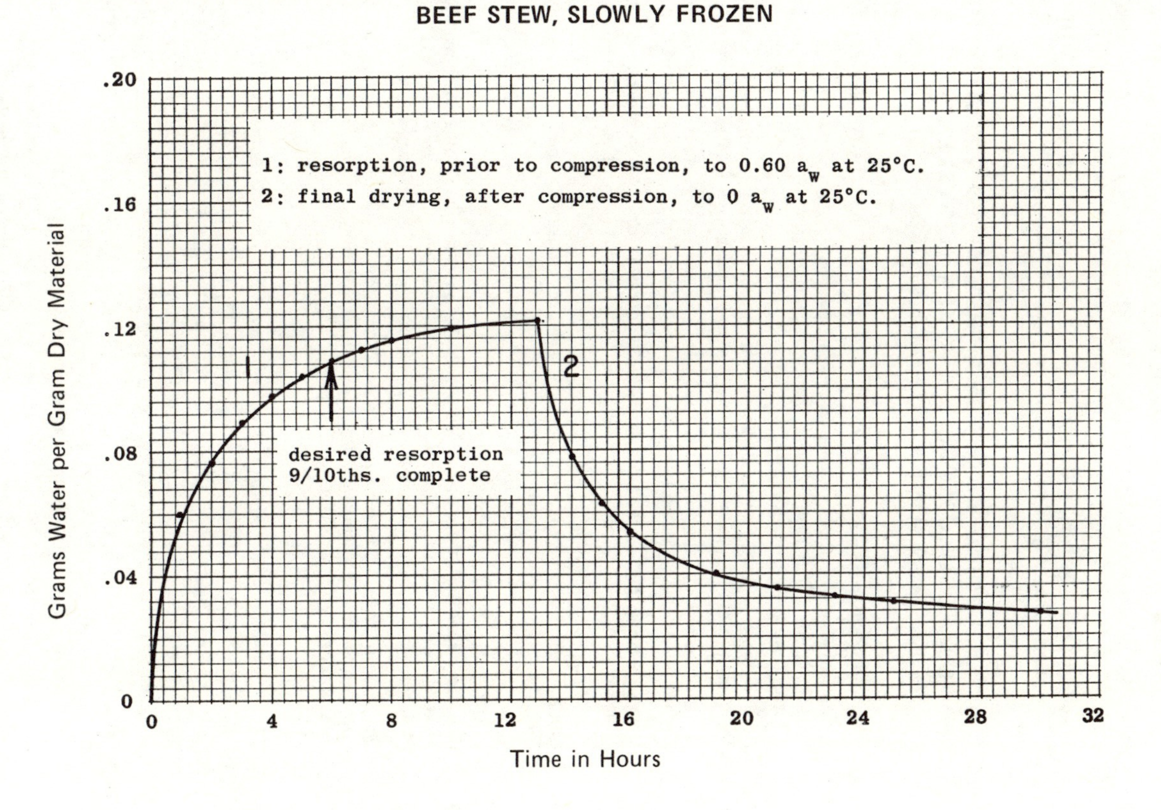

Source: Technical Report 72-33-FL, December 1971

Restoration of Compressed, Solvent-Extracted Material

The results of physical compression/restoration studies conducted on various solvent-extracted materials are summarized in Table 1.10. Behavior is listed with reference to water activity prior to compression. Slowly frozen, freeze-dried, solvent-extracted cooked beef recovered best when adjusted to 0.60 a_w prior to compression. On the basis of earlier observations, the best a_w for solvent-extracted, cooked beef would appear to be 0.15 units higher than that for control cooked, freeze-dried beef.

TABLE 1.10: EFFECTS OF WATER ON FREEZE-DRIED AND SOLVENT-EXTRACTED FOODS COMPRESSED AFTER REMOISTENING TO VARIOUS WATER ACTIVITIES

a_w	Beef (Cooked)	Chicken (Cooked)	Apple (Fresh)	Pineapple (Canned)
0.80	Compresses very readily; poor restoration.	Compresses very readily; poor restoration.	Compresses very readily; fair restoration.	Compresses very readily; poor restoration.
0.70	Compresses readily; fair restoration.	Compresses readily; good restoration.	Compresses readily; good restoration.	Compresses readily; fair restoration.
0.60	Compresses readily; good restoration.	Compresses readily; fair restoration.	Compresses readily; fair restoration.	Compresses readily; poor restoration.
0.50	Compresses readily; fair restoration.	Compresses readily; some breakage; fair restoration.	Compresses readily; poor restoration.	Compresses readily; poor restoration.
0.40	Compresses readily; fair restoration.	Shatters on compression; pieces show fair restoration.	Compresses readily; very poor restoration.	Compresses readily; very poor restoration.
0.30	Compresses with some breakage; fair restoration.	Crumbles on compression; no restoration.	Compresses readily; very poor restoration.	Compresses readily; very poor restoration.
0.20	Shatters on compression; no restoration.	Powders on compression; no restoration.	Compresses readily; no restoration.	Compresses readily; no restoration.

Cooked chicken subjected to slow freezing, freeze-drying and solvent-extraction recovered best when resorbed to 0.70 a_w prior to compression, that is, to a water activity 0.10 units higher than that found best for cooked freeze-dried material not subjected to lipid extraction (Tables 1.10 and 1.2, respectively). Apple extracted with aqueous ethanol recovered best when exposed to 0.70 a_w prior to compression, less well at 0.60 and 0.80, and very poorly at 0.50 and lower values. Such performance stands in marked contrast to that of fresh apple tissue which, when frozen and freeze-dried, recovered best when exposed to 0.30 a_w prior to compression. Solvent-extracted pineapple made a somewhat less impressive recovery than apple. Much as in apple, however, a large change in best a_w was observed to accompany solvent extraction. Best a_w was, that is, raised from 0.25 to 0.70.

Cytological Studies

Cooked Beef: Slowly frozen tissues subjected to freeze-drying and rehydration but not to compression differed very little in appearance from cooked controls. Fibers failed, however, to regain their initial cross-sectional areas. Freeze-drying appeared, therefore, to reduce the water-holding capacity of the individual fibers. Greatest damage was observed on rehydration when compression followed exposure to 0.20 a_W. Fibers were fractured and displaced. Little obvious damage was detected in tissues exposed to 0.50 a_W prior to compression. In specimens resorbed to 0.80 a_W, compressed and rehydrated, fibers were aggregated apparently from fiber-to-fiber adhesion resulting from compression. In parallel studies, rapid freezing prior to freeze-drying was found to permit still better recoveries. With the exception of the specimens compressed at 0.20 a_W the structures on rehydration were hardly distinguished from the cooked controls.

Cooked Chicken: Exposure to freezing, freeze-drying and rehydration altered the appearance of the fiber contents but did not result in physical damage. Compression at 0.40 a_W resulted in considerable damage; neither fibers nor connective tissues regained their initial form on rehydration. Compression at 0.60 a_W resulted in reduced damage; compression at 0.80 a_W seemed to be still less harmful. Best behavior was not, therefore, correlated with an intermediate a_W though water activities were chosen after completion of other studies. In a corresponding series of observations on cooked chicken subjected to rapid freezing, the freeze-dried control and the samples exposed to 0.40 and 0.60 a_W prior to compression recovered somewhat better than did the slowly frozen ones. Rapidly frozen samples compressed at 0.80 a_W recovered less well.

Canned Tuna: The greatest observable changes in tuna appeared to result from freezing and freeze-drying. Fibers were markedly condensed; they lost, in part, their ability to recover, i.e., to hydrate to the original level and acquired a brittle quality. Connective tissues were observed to form a striated coagulum. Compressed materials exhibited better recovery the higher the a_W prior to compression. On the basis of appearance only the damage due to compression may be summarized: damage (0.40 a_W) $>$ damage (0.60 a_W) $>$ damage (0.80 a_W). Rapidly frozen materials were not examined.

Apples: Very little damage was immediately apparent in any specimen. Cellular arrangements were excellently retained in materials subjected to freeze-drying and to rehydration and in rehydrated specimens exposed to 0.40 a_W prior to compression. Further inspection, however, revealed a near absence of intercellular spaces in all specimens subjected to compression. Breaks in the cell wall appeared here and there in 0.20 and 0.60 a_W compressed materials. Greater tendencies to recover (or to assume) rounded shapes were noted in materials exposed to 0.60 a_W prior to compression. Elongated cells suggestive of flattened structures were seen to persist more frequently in specimens compressed at 0.20 and 0.40 a_W. None of the compressed, restored specimens was, however, distorted in the plane perpendicular to the direction of compression.

Pineapple: The canning process seemed not to affect the structure of the tissues save perhaps to cause cell walls to break occasionally. Freezing and freeze-drying resulted in more extensive damage. Cell walls were broken in greater numbers. A partial collapse resulting from freeze-drying was, moreover, not completely reversed by rehydration. Compression/restoration treatments did not seem to contribute to further damage at 0.25 a_W. Tissues exposed to 0.40 a_W prior to compression were, however, found to retain impressed configurations upon rehydration.

DISCUSSION

The several separate studies reported have been grouped for the purposes of discussion into three categories. Thus, the significance of the results will be considered with reference (a) to the small scale production experiments, (b) to the pilot scale practice, and (c) to the theory of the processes related to compression.

Compression/Restoration Behavior of One-Component Systems

An analysis of the results of the laboratory compression/restoration tests and a comparison of the latter with the verdicts returned by the taste panel require discussion with reference to water content at time of compression. The properties of the various single foods are thus examined with reference to the relevant sorption isotherms as follows.

Apples: Best restoration was observed in apples compressed after resorption to 0.30 a_W. Thus, from the resorption isotherm, the water content for best restoration was 0.050 g/g dry tissue. Such a value compares with 0.045 g/g for peach and 0.035 to 0.050 for pineapple.

These best water contents are so much lower than those required in plant tissues less rich in sugar and in other foods that it seems worth the while to consider the possible cause of the differences. To a first approximation the freeze-dried plant tissue can be viewed as a two-phase, two-component system composed of cellulose and sugar. On exposure to water the sugar phase, doubtless amorphous after freeze-drying, sorbs more water than the cellulose. Given a high enough a_W, the sugar phase softens and flows under stress, the cellulosic phase deforming simultaneously, with reluctance.

Since the sugar-rich phase flows only above its glass transition temperature, the sugar/water ratio has to be decreased to cause the latter to fall below the temperature selected for compression. Where the system is moistened further, the sugar-rich phase may flow too readily, the various internal surfaces being annihilated in the process. The cellulosic component may, moreover, lose its ability to retain a strained form at high water activities, the sugar and the excess water plasticizing the irreversible deformation. The restoration observed where apples were placed in hot water tends to support such a hypothesis. Irreversible deformation might prove reversible where sufficient energy (i.e. heat) was available.

The failure of the thicker slabs where single pieces gave a good performance calls for close attention. Evidently a way must be found to permit water to reach each piece in the slab. Similar problems arose in several other foods in this study.

Chicken: Good restoration was obtained from slowly frozen chicken compressed after direct desorption, also after resorption from dryness. That is, recovery was not dependent on the avoidance of or the passage through certain states of dryness prior to compression. The good recovery demonstrated by chicken frozen slowly prior to limited freeze-drying and to compression may, however, be contrasted with the poor recovery shown by chicken compressed after rapid freezing and limited freeze-drying. Clearly, the freezing rate was a most important factor.

The strong dependence of the best recovery of the slowly frozen chicken on water activity prior to compression may be traced, through the appropriate isotherm, to dependence on water content. Both desorption and resorption isotherms are seen to be very steep in the regions in which best behavior was observed; that is, best restoration was very strongly dependent on water content.

It is, however, also quite clear that the water content best after limited freeze-drying (0.140 g/g dry tissue) differed from that required (0.105 g/g) when dry material was resorbed at 25°C. It would appear that, where the protein has been exposed to drying, a lesser *quantity* of water suffices to render the tissue deformable. One must suppose that the water bound during resorption does not have access to the sites occupied during desorption or, at least, that the sequence according to which the sites are occupied is not obtained by reversing the sequence in which the sites were vacated during desorption.

The taste panel judged desorbed materials compressed at 0.50, 0.60 and 0.70 a_w equally good in each case. The panels' opinions thus coincide, inasmuch as the mean value proves to be 0.60, with the results of the physical tests. The panels' observation that the rapidly frozen material restored less well is also most interesting. It is perhaps especially interesting that the rapidly frozen material restored at all when dried and resorbed when it failed (during physical tests) to restore when prepared by limited freeze-drying.

Cottage Cheese: Equally good recovery by slowly frozen, resorbed and rapidly frozen, desorbed dry curd cottage cheese contrasts with the fair recovery demonstrated by the slowly frozen desorbed product. No other food was found to offer better restoration after rapid freezing. Possibly these observations reflect the granular nature of the cottage cheese and the disorder characteristic of the denatured protein present.

From the desorption and the resorption isotherms obtained from the dry curd product, the strong dependence of the best behavior on a_w can be traced to best water contents in the range 0.10 to 0.12 g/g dry material regardless of the freezing rate or the form of the subsequent treatment. Apparently the water molecules are not distributed according to a pattern determined by the state from which the moist condition was approached. Further clues to the mechanism may be found in the observation that the dry curd and the creamed cottage cheese products, each slowly frozen, freeze-dried and resorbed, restored best at the same a_w. It would appear, in this connection, that the cream does not reside in places critical to compression or to restoration. The problems involved are therefore seemingly related to the behavior of the proteins present.

Mushrooms: The absence of a really adequate performance at any water activity would appear to originate in the rather elaborate form of the fresh material. In addition to enclosing in part a series of empty spaces, fresh mushrooms contain considerable quantities of air. Proteins, sugars, minerals and fibers are present in the ratio, roughly, of 3:4:1:1.

The best recovery, obtained at 0.50 a_w, requires, according to the resorption isotherm, a water content of 0.050 g/g dry tissue. Probably the water acts in two ways: first to soften the sugar/mineral glass, second to soften the protein and fiber matrix. Most likely, compression results in extensive internal surface-to-surface adhesion not reversed on rehydration. Possibly the incorporation of an additive to prevent direct surface-to-surface contact of the various parts of the mushroom would help. Canned mushrooms, not examined in this study, also deserve serious consideration. Their obvious resilience appears, in retrospect, to their potential advantage.

Noodles: When the best water contents are determined from the respective sorption isotherms, one sees the resorbed material restores best when compressed at water contents much higher than those making possible best performance by direct desorption. In contrast, the freezing rate prior to freeze-drying does not influence the behavior. According to expectation, the addition of glycerol was observed to lower the best water activity. Glycerol was also observed to broaden the range of useful water activities. In the absence of additional determinations, it is not known whether glycerol lowers the best water content.

Several factors require further examination. Tendencies to regain volume and texture might be distinguished from tendencies for piece-to-piece adhesion. The first and second factors are most likely susceptible to considerable manipulation on the basis of existing knowledge of the properties of starch gels and the effects of different thermal treatments in various sequences. The problem of adhesion is perhaps better discussed with reference to the possible effects of additives and/or edible coatings.

Pineapple: The very good recovery observed in hot water contrasts with its near absence in cold water. While it is still impossible to distinguish the reasons for the marked difference in behavior, the poor performance in cold water is very likely related to the adhesion of the cell walls, one to another. The strongly a_w dependent behavior in hot water recalls the similarly strong dependence where apples were restored in cold water. The resorption isotherm indicates a best water content of 0.040 to 0.050 g/g dry tissue, suggesting as it did in apples, the mediation of a concentrated aqueous sugar solution during compression, the cellulosic component being but slightly hydrated. The rating accorded frozen material resorbed at 0.25 and 0.30 a_W suggests the potential of the pineapple *as long as* restoration is induced in hot water. The scores further demonstrate the lack of adverse effect resulting from brief exposure to high temperatures. Clearly the slowly frozen pineapple was difficult to damage!

Rapid freezing was, in comparison, very deleterious. None of the rapidly frozen compressed material prepared for the panel was restored after 30 minutes in hot water. Since rapidly frozen controls were readily rehydrated, one concludes that the rapid freezing *in combination with* the compression was harmful. In the absence of any positive correlation (a) of rapid cooling with intracellular freezing or (b) of slow freezing with intercellular ice formation, further discussion is beside the point.

Similarly, pineapple collapsed during limited freeze-drying was not restored, indicating perhaps very extensive self-inflicted compression. Moderately low temperatures, high water activities and long times of exposure appeared to act together to cause the total disappearance of the channels created during freezing, possibly also the accommodation of the cellular components in new strain-free states.

Potatoes: While partial restoration was obtained with potatoes compressed at water activities in the range 0.50 to 0.70, that is, at water contents from 0.060 to 0.160 g/g, difficulties encountered demand special discussion.

Crumbling during compression, attributed to a too-dry state, was observed to extend to water contents high enough to permit an irreversible adhesion generally attributed to a too-moist state. One problem was not overcome, that is, with change in water content, before another made its appearance. Additives employed in low concentration in processed, sliced potatoes resulted in only moderately improved behavior. Probably the crumbling persisted in the starch at water activities in which cell-to-cell adhesions were already problematical. Satisfactory solutions might be found in the selection of especially young potatoes. Alternatively, cooked potatoes subjected to a suitable thermal treatment prior to freezing might acquire a sufficiently durable structure. Methods might have to be devised to permit the effective introduction of small additive molecules into freshly sliced materials.

Tuna: Most remarkable, perhaps, were the widths of the ranges within which desorbed and resorbed tuna demonstrated good recovery with reference to water activity. Brief inspection of the relevant sorption isotherms shows correspondingly wide ranges in permissible water content (from 0.110 to 0.220 g/g dry tissue) independent, more or less, of the way in which the water contents were achieved. Equally remarkably good recoveries were in fact obtained from water contents where tissues cracked when compressed one piece at a time. Restoring forces released in tuna during rehydration must be very strong.

Somewhat less acceptable restoration resulting from rapid freezing seems to follow the pattern observed in chicken. In tuna, as in chicken, rehydration would appear to require the opening of channels in the protein phase. Greater internal surface areas resulting from the freeze-drying of the rapidly frozen material were most likely eliminated more readily during compression than those, fewer, larger cavities resulting from slow freezing. Smaller spaces, once destroyed, were furthermore less readily redeveloped.

In Summary: One could question the need to discuss compression/restoration behavior with reference to water activity. Brief consideration of the foregoing analysis of the behavior of single food products shows, however, the usefulness and the limitations of the water activity approach. Clearly, best behavior was defined in each case in terms of a_w in the absence of any knowledge of the water contents involved. Preparation for compression was, moreover, conducted in equipment in which a_w's were, one way or another, predetermined.

In contrast, a knowledge of best water content permits preparation for compression only by direct admixture, one way or another, of food product and water. With a knowledge of the best a_w and the sorption characteristics (i.e., the isotherm) one is free to select any preparative method. Notwithstanding these alternatives, it is the composition that appears very largely to determine the physical properties. Thus, a knowledge of water contents is essential (either way) to any discussion of the molecular basis for the best behavior. Desorption isotherms appear to be of particular value inasmuch as direct desorption to predetermined water contents at any of a variety of freeze-drying temperatures may prove desirable.

Compression/Restoration Behavior of Mixed Food Products

Interesting problems arise in mixed foods where each component exhibits best performance at an a_w and a water content characteristic of that component. Further interest attaches to the modification of the best water activities where substances originating in one foodstuff penetrate another. Similarly the effects of various added substances command attention. The two mixed food products examined in the course of this study are, as far as possible, examined from these points of view.

Beef Stew: It is of special interest that the best performance of the five component mixture was determined by physical test and by the taste panel to depend less on a_w prior to compression than did the several components in separate tests. If these judgments reflect accurate objective analyses, it is likely that, in the processing of the mixture, each component is affected by the presence of others to the benefit of the respective components' abilities to recover. If, on the other hand, improved acceptance results from decreased discrimination by the panel, the benefits attached to the preparation of mixed food products are not to be minimized. Additional studies will be required before the effects of the substances released by one component on the performance of another can be determined. Tendencies for certain component surfaces to interact and adhere should also be examined in greater detail.

Tuna Casserole: While the physical tests showed the recovery to be excellent at 0.60 and 0.70 a_w, the taste panel judged the materials compressed at 0.50 and 0.60 a_w to be preferable to those compressed at 0.70 a_w (an informal assessment despite the absence of a true statistically significant difference). Probably the difference in best a_w relates to the difficulties encountered in the rehydration of the larger slabs compressed at the higher a_w's. Adhesion seems to be reversed less easily, the larger the sample. Partial fragmentation resulting from compression at lower a_w aids, in contrast, in the creation of the pathways by which water penetrates the larger samples.

Since the tests showed the various components to recover as well (but no better) in the presence of each other than in their absence, further discussion can, it appears, be conducted with reference to the resorption isotherms obtained from the various separate components. The mushrooms are found to exhibit the lowest optimum a_w and the greatest a_w dependence, that is the steepest isotherm. A more compatible mixed food product would result, were it possible to lower the best a_w's of the noodles and the tuna (the freeze-dried white sauce matrix would in the process be less likely to soften and collapse). A search for ways to raise the best a_w for the mushrooms, or to flatten the mushroom isotherm, would, on the other hand, perhaps serve equally well. It is sufficient at this stage to point out the various possibilities.

In Summary: It is evident that foods can be matched with reference to the dependence of the properties of each on a_w to permit the formulation of mixtures capable of uniformly good recovery after compression. Inasmuch as consumers appear less critical, perhaps, of the performance of components in admixture than of the same foods taken one at a time, the matching process is facilitated.

Sources of water of predetermined a_w were shown to provide effective means of preparing mixed freeze-dried materials for compression. From the desorption isotherms and the composition of the mixture, the weight of water resorbed could be calculated in advance where such information was desired. The resorption isotherms describing the behavior of the single foods permit the determination of the components most usefully retained, eliminated or modified. The modification of the properties of individual components appears, in particular, to offer a rather promising means of obtaining (or improving) useful mixtures.

Rate Determinations

Rates: The rate determinations would appear to contribute useful information. Specifically the freeze-drying rates helped to distinguish the food products less willing to freeze-dry—less willing, that is, to permit the diffusion of water (since the heat transfer was, it would appear, about equally good in each experiment).

The measurements of the rates at which materials resorbed water, via the vapor phase at 25°C, demonstrated the practicality of moistening by resorption. The plant tissues rich in sugar resorbed much less rapidly than the protein foods. The resorption measurements very likely indicate also the upper limits of rates of equilibration where liquid water is added directly to food (the water must still penetrate each part of each component). A discussion of other factors determining the distribution of the water added directly to freeze-dried foods is, however, not within the scope of this report. The measurements of the rates of final desorption indicate the times required for the final drying of slabs several millimeters thick. The finite nature of the rates in question should not be discounted. The freeze-drying and resorption rates were perhaps more dependent on the experimental geometry than the desorption rates. The former involved large changes in the temperature of the sample in comparison with sample chamber/condenser and sample chamber/evaporator temperature differences, respectively. Specimens cooled during final desorption but little in comparison with sample chamber/condenser temperature differences.

Variation in Freezing Rate Prior to Compression

Cooked Beef: The decreased restoration exhibited by the beef frozen in liquid nitrogen suggests failure of the water to penetrate the collapsed labyrinth within the fibers. (Previous studies have shown the highest of the freezing rates employed to cause a growth of hundreds of separate ice crystals inside each fiber.) Since, during compression, the physical displacements within the fibrillar matrix are clearly less extensive the higher the freezing rate, the failure to rehydrate is, perhaps, best attributed to tendencies on the part of the intracellular surfaces to mutual, irreversible adhesion.

Carrots: The lesser resistance to compression detected in the more rapidly frozen carrots can best be explained in terms of freezing-freeze-drying hysteresis. The large distortion introduced during freezing very likely results in a physically stronger matrix, softening on resorption notwithstanding. A reduced recovery detected in the carrots frozen at the higher velocities (a marginal difference, to be sure) could be linked, as in beef, to greater tendencies to adhesion, one surface to another, within the compressed structure. The greater resistance to compression accompanying the slower freezing most likely reduces the incidence of the surface-to-surface adhesion. The problem seems to be compounded by rapid freezing in that it results (a) in greater internal surface area and (b) states from which the surfaces thus created are more likely to meet.

Chicken: The complete recovery of the rapidly frozen chicken despite the better compression can only be explained in a manner consistent with the previous arguments in terms of a reversible adhesion (cf beef). Possibly the better behavior is associated with the lower lipid content or with ultrastructural differences. A reduction in the rate of recovery with increased freezing rate is presumably the direct result of the subdivision of the routes available for the reentry of water.

Solvent Extraction

Beef: Table 1.11 summarizes the combined description of whole foods and of the corresponding solvent-extracted materials. This work confirms that the best a_w was raised 0.15 water activity units by solvent extraction. It would seem that the lipids do contribute to the adhesion of the component surfaces, one to another, within the compressed material. In their absence, the supposed adhesion requires a more extensive moistening (and softening) of the protein.

Chicken: An increase in best a_w with solvent extraction somewhat less marked than that in beef correlates with the known low lipid level in light chicken muscle. Corresponding studies on dark meat would seem to be indicated.

Apples: Table 1.11 shows the very extensive shifts in best a_w resulting from the extraction of the sugar-rich plant tissues. Evidently apples behave much the same way as peaches. Most likely the sugars bind sufficient water at low a_w's to plasticize the cellular matrix and, at higher a_w's, to permit the total elimination of the spaces by which water might return. The cellulosic structures themselves are, by contrast, much less prone to bind water. Only at 0.7 a_w do they appear to soften sufficiently to deform without physical rupture.

Pineapple: While the best performance after solvent extraction was never better than fair, the best a_w was shifted to a higher water activity than it was in apple; that is, the best a_w was raised very considerably. Probably the same arguments apply. The plasticizing effects of the sugar solutions developed when the glass-like regions resorbed water would seem to claim a place in any discussion of the restoration of compressed plant tissues.

TABLE 1.11: COMPARISON OF THE PERFORMANCE ON REHYDRATION OF WHOLE AND SOLVENT-EXTRACTED COMPRESSED, FREEZE-DRIED TISSUES

a_w:	0.20	0.30	0.40	0.50	0.60	0.70
Beef, cooked, slowly frozen, whole:	P	F	G	G	F	P
Beef, cooked, slowly frozen, extracted:	P	F	F	F	G	F
Beef, cooked, slowly frozen, extracted:	–	F	F	F	F	F
Beef, cooked, rapidly frozen, extracted:	–	F	F	G	G	G

a_w:	0.30	0.40	0.50	0.60	0.70	0.80
Chicken, cooked, slowly frozen, whole:	P	P	F	G	F	P
Chicken, cooked, slowly frozen, extracted:	P	F	F	F	G	P

a_w:	0.20	0.30	0.40	0.50	0.60	0.70	0.80
Apple, fresh, slowly frozen, whole:	F	G	F	F	–	–	–
Apple, fresh, slowly frozen, extracted:	P	P	P	P	F	G	F

a_w:	0.20	0.25	0.30	0.40	0.50	0.60	0.70	0.80
Pineapple, canned, water pack, slowly frozen, whole:	F	G	F	P	P	–	–	–
Pineapple, canned, water pack, slowly frozen, extracted:	P	–	P	P	P	P	F	P

(G = good F = fair P = poor)

Table 1.12 lists the fiber and the sugar contents of various plant tissues. Among the tissues richer in sugars, the best recovery (admittedly in hot water) was observed in the tissue lowest in fiber content, having the highest sugar/fiber ratio. Excellent recovery was, however, obtained from the cooked carrots having the highest fiber content and the lowest sugar/fiber ratio. Much remains to be learned.

TABLE 1.12

Food	Sugar	Fiber	Ratio	Recovery	Best a_w
Apple, fresh	13.5	0.6	22:1	Good	0.30
Peach, fresh	8.5	0.6	14:1	Very poor	0.30
Pineapple, canned	9.9	0.3	33:1	Excellent	0.25
Cabbage, fresh	4.6	0.8	6:1	Very poor	0.35
Carrot, cooked	6.1	1.0	6:1	Excellent	0.55

Cytological Studies

Beef, Cooked: Cytological studies support in some respects the findings of physical compression/restoration studies. Evidence of fiber breakage resulting from compression was observed in the restored specimens compressed at 0.20 a_w and, in small measure, in those compressed at 0.50 a_w. While no fiber breakage was seen in the specimens compressed at 0.80 a_w, fiber aggregation was detected; that is, fibers brought into contact failed to separate upon rehydration. A decreased water binding resulting from freeze-drying (in specimens not compressed at all) was also detected. Improved compression/restoration behavior thus depends to some extent upon improved response to freeze-drying and/or to final drying. (It would not be difficult to distinguish the separate effects of the two last named treatments.) Possibly, final drying could be limited to a reduction to a water activity (hence to a water content) measurably greater than zero.

Chicken: In contrast with beef, chicken control specimens, freeze-dried and rehydrated, appeared to regain their original size and form; fibers were equally large; spaces between fibers were correspondingly small. Such observations correlated well with the taste panels' informal opinions of the respective control materials. The steadily increased performance of the compressed, rehydrated chicken with increased a_w did not correlate with the observation that the chicken recovered best when compressed at 0.60 a_w. The very long times during which the tissues were rehydrated and fixed prior to transfer to paraffin wax may, however, have permitted a time-dependent restoration not noticed in tests limited to 30 minutes, or thereabouts. These observations call into question such terms as irreversible damage.

Tuna: The lack of any sharp dependence of good recovery on a_w prior to compression is borne out by the microscopic study. In this respect, tuna is shown to be the least a_w dependent muscle tissue, chicken being more dependent and beef much more so. Presumably the tuna muscle tissue is less brittle at lower

a_w, less adhesive at higher a_w. In the absence of a detailed comparative study, one can only speculate as to the cause of the different behavior. Possibly the behavior of the tuna is to be traced to the extensive distribution of nonfibrillar material, this being the most obvious of the differences visible under the microscope.

Apples: The lack of obvious damage to the restored apple tissues is of little direct help in determining the causes of the failure to restore. The micrographs do not, however, indicate the absence of a turgor pressure, nor do they necessarily indicate the original structure prior to fixation. Probably the prolonged exposure to water and afterwards to fixative, results in restoration in excess of that achieved in a shorter test. The absence of the extracellular spaces in the compressed, rehydrated tissues depends, presumably, on the nature of the surfaces turned toward the spaces. Inasmuch as a system composed of such interconnecting spaces might constitute a natural route to rehydration, the nature of the cell's surface is of considerable interest.

Pineapple: Despite the suggestion that the freeze-drying itself results in some damage, the pineapple demonstrates a remarkable power to recover in hot water. The vascular bundles recover especially well, together with the cells surrounding them. The tendency on the part of the flimsy cellulosic framework to resume its original size and shape in the presence of so much sugar could result from a resilience inherent in the fiber structure. A tendency for water to enter damaged cells at a rate many times that at which sugars could diffuse away would, however, have the same effect. Doubtless the absence of an osmotic response could be demonstrated.

CONCLUSIONS

1. Best recoveries are most usefully defined in terms of water activities to which separate foodstuffs are adjusted prior to compression.

2. Foods can, in most cases, be prepared for compression by direct desorption or by remoistening from a too-dry state with equal ease and effectiveness. Freeze-dried foods appear to desorb and/or resorb water via the vapor very rapidly.

3. Knowledge of the relevant sorption isotherm permits the determination of the best water content prior to compression, corresponding to the best a_w.

4. Rational choice of foods destined for compression in admixture can only be made on the basis of water activity, compression at which yields best recovery in the case of each projected component.

5. There is little or no reason to doubt that information gained from the small scale experiments presented here cannot be used as a basis for large scale operations.

6. Freeze-drying by sublimation of ice and direct desorption of a portion of the water remaining unfrozen (limited freeze-drying) appears to offer certain practical advantages. The method seems to be well-suited to pilot scale studies in certain instances.

7. The freezing rate appears in some cases to determine the recovery of the freeze-dried compressed material. Foods high in fats or sugars recover better when first frozen slowly. Foods rich in protein but devoid of fat seem to recover better when frozen rapidly.

8. Physical damage to structural components appears to determine failure to recover from compression at too-low water activities.

9. Lipids are rather clearly involved where freeze-dried muscle tissues fail to recover from compression at higher water activities.

10. Sugar content per se does not seem to determine extent of best performance of freeze-dried, compressed plant tissues. Rather, the failure to restore appears to be related to the nature of substances of higher molecular weight.

REFERENCES

(1) Jensen, W.A. (1962), *Botanical Histochemistry*, W.H. Freeman and Company, San Francisco, California, pp. 90-94.
(2) MacKenzie, Alan P. (1964), *Biodynamica*, 9, pp. 213-222.
(3) MacKenzie, Alan P. (1965), *Bulletin of the Parenteral Drug Association*, 20, pp. 101-129.
(4) MacKenzie, A.P. (1967), *Cryobiology*, 3, p. 387 (abstract).
(5) Mason, B.J. (1957), *The Physics of Clouds*, Clarendon Press, Oxford, p. 445.
(6) Sass, J.E. (1964), *Botanical Microtechnique*, Iowa State University Press, Ames, Iowa, pp. 55-77.
(7) Watt, B.K., and A.L. Merrill (1963), *Composition of Foods*, United States Department of Agriculture, Washington D.C.

The Effect of Compression on The Texture of Dehydrated Vegetables

The material in this chapter is drawn from two reports from the U.S. Army Natick Laboratories. The first, Technical Report 70-36-FL, entitled *Studies on Reversible Compression of Dehydrated Vegetables,* was done by A.R. Rahman, Glenn Schafer, G.R. Taylor, and D.E. Westcott, and is dated November 1969. The second report, *Increased Density for Military Foods,* was written by A.R. Rahman, W.L. Henning, Sheldon Bishov, and D.E. Westcott in 1972.

These studies were made to attempt to define optimum processing conditions for achieving the compression of freeze-dried foods without sacrificing quality and acceptability, and to attempt to define the effects of pressures up to 2000 psi on the cell structures and textural qualities of freeze-dried foods. This chapter discusses reversibly compressed vegetables which are designed for consumption only in the reconstituted state.

REVIEW OF PAST WORK

The feasibility of compressing foods and subsequently restoring them was first recognized by Ishler (1). He indicated that successfully compressed freeze-dried cellular foods can be achieved by spraying with water to 5-13% moisture, compressing and redrying to less than 3% moisture.

These findings were later confirmed by Brockmann (2). Later techniques were remarkable improved, the most notable studies being those of Mackenzie (3) and Rahman (4) relative to conditioning, compression pressure and dwell time procedures.

However, the texture (turgidity) of rehydrated products after compression is usually less firm than their counterparts due to the numerous treatments required before the compression of foods, namely blanching, freezing and freeze-drying. The physical properties that reflect the turgidity depend largely upon the structural arrangement and chemical composition of the cell walls. Since the intercellular cement is primarily composed of pectic substances (4)(5)(6) any agent or process which breaks down those substances can obviously bring about cell separation. Heating can cause breakdown of pectic substances in fruits and vegetables and ultimate separation of intact cells (7)(8)(9). Further processing such as freezing and freeze-drying can cause further changes to the cells. This was realized by Fennema (9) and Reeves (10) who indicated that freezing, especially slow freezing of fruit and other multicellular structures, often damages the tissue and ultimately the texture.

EXPERIMENTAL PROCEDURES

Product Preparation

All products were locally purchased. Compression was accomplished on a Carver press using compression forces of 500, 1000, 1500, 2000, and 2500 pounds per square inch (psi). After the compression, all the compressed bars were redried using a vacuum oven to reduce the final moisture content to approximately 2%.

Peas: Individually quick frozen (IQF) peas were partially thawed and the seed coat was mechanically slit at several points in order to facilitate the removal of water during freeze-drying. The peas were sulfited by dipping in solution of sodium metabisulfite to yield approximately 400 ppm. They were frozen at -20°F and then freeze-dried at a platen temperature of 120°F to a final moisture content of less than 2%. The freeze-dried peas were subjected to live steam for 5 minutes. 20 grams of peas were compressed to bars about 3 x 1 x ½ inches.

Corn: Corn was freeze-dried with a platen temperature of 120°F to a final moisture content of less than 2%. The freeze-dried corn was sprayed with water until the weight was increased by 12%. The sprayed corn was then heated in a closed container for 10 minutes at 200°F to equilibrate the moisture in the corn. The corn was then compressed.

Spinach: Frozen spinach was thawed and sulfited by dipping in a solution of metabisulfite to yield approximately 500 ppm. It was then freeze-dried at a platen temperature of 120°F for 16 hours. The freeze-dried spinach was subjected to live steam for 5 minutes and compressed. Each bar weighed about 10 grams.

Onions: Sliced onions were air dried as in Spec JJJ-0-533 (11), treated with live steam for 5 minutes and compressed.

Green Beans: Cross cut frozen green beans were freeze-dried and compressed following the procedure for the spinach.

Carrots: Fresh carrots were lye peeled and cut into ⅜ x ⅜ x $^{1}/_{16}$ inches. The carrots were steam blanched for 10 minutes in order to inactivate the enzyme, peroxidase. The carrots were then sulfited by dipping in a solution of metabisulfite to yield approximately 400 ppm. They were then frozen at -20°F and freeze-dried with a platen temperature of 120°F for 16 hours. They were then compressed following the procedure used for the spinach.

Rehydration Ratio

Rehydration ratio was determined by dividing the rehydrated weight by the dry weight. The following rehydration procedures were used.

Corn and Peas: Each bar was placed in approximately 500 ml of 210°F water, held for 12 minutes and then drained for 5 minutes.

Carrots: Each bar was placed in approximately 500 ml of boiling water, boiled for 1 minute and soaked for another minute and then drained for 5 minutes.

Onions: Each bar was placed in approximately 500 ml of boiling water, boiled for 5 minutes, and then drained for 5 minutes.

Spinach: Each bar was placed in approximately 500 ml of water at 210°F, held for 3 minutes, and then drained for 5 minutes.

Green Beans: Each bar was placed in approximately 500 ml of water at 210°F, held for 15 minutes and then drained for 5 minutes.

Test Procedures

Compression Ratio: To determine compression ratios, the dehydrated vegetables were compressed at 1500 psi into discs approximately 3¾ inches diameter, so that they would fit into a number 2½ can. The compressed discs required to fill the can, leaving approximately ¼ inch headspace, were weighed. Uncompressed freeze-dried product of equivalent weight to that of the compressed was packed loosely in number 2½ cans leaving approximately ¼ inch headspace. The number of cans utilized to pack the loose product gave the compression or packaging ratio.

Bulk Density: Bulk density was measured by dividing the weight of product by its respective volume to yield grams per cubic centimeter. Calculated compression ratio was then determined by dividing rhe bulk density of the compressed product by that of the uncompressed.

Shedding Time: This was measured by placing the compressed product in hot water, approximately 210°F, and then reporting the time at which all the individual pieces separate from the compressed product.

Texture: Texture was measured with the Allo-Kramer Shear Press using the 5000 pound ring with a 30 sec downstroke. In addition, texture was evaluated by technological taste panel of 10 trained judges using 9 point scale ranging from 1 = extremely poor to 9 = excellent (15).

RESULTS

Results indicate that compressed foods can be successfully produced without appreciably sacrificing their overall quality. Reduction in volume ranging from 4 to 16 fold have been achieved. For example, one truckload of freeze-dried, compressed green beans contains as much food as 16 truckloads of the uncompressed. This is extremely significant in terms of savings in packaging material and logistical advantages of reduced handling, storage and transportation. In addition, environmental improvement due to a lowered waste disposal requirement will result.

Tables 2.1 through 2.6 give the texture and rehydration ratio for each of the compressed vegetables. These data indicate that compression forces used did not produce significant differences in rehydration ratio or texture as measured by the shear press. However shedding time increased somewhat as the compression force increased. This suggests that shedding time may reflect the degree of cohesiveness as a criterion for the ability of compressed products to withstand abuse during handling, storage and transportation.

Table 2.7 shows that neither conditioning moisture level nor compression force affected the rehydration ratio and the texture of compressed peas as measured by the shear press or by a technological panel. Histological studies showed no apparent difference in cell separation or disruption of cell walls in peas compressed at 1,000, 1,500 or 2,000 psi compared to uncompressed peas. Results shown in table 2.8 indicate that compression ratio as determined by actual filling of cans, is slightly lower that that calculated from the bulk densities. This is due to the allowances given to headspace, the space between the compressed discs and the can wall needed to facilitate packing and unpacking of the product, and space between the discs due to uneven surfaces caused by relaxation of the product after compression.

TABLE 2.1: TEXTURE AND REHYDRATION RATIO OF DEHYDRATED PEAS AS AFFECTED BY COMPRESSION

Compression Pressure, psi	Rehydration Ratio*	Shear Press Value After Rehydration, lb*	Shedding Time, min
500	4.26	279	1
1,000	4.46	283	3
1,500	4.24	287	3
2,000	4.30	309	4
2,500	4.69	309	4¼

*No significant difference at 5% level.

TABLE 2.2: TEXTURE AND REHYDRATION RATIO OF DEHYDRATED CORN AS AFFECTED BY COMPRESSION

Compression Pressure, psi	Rehydration Ratio*	Shear Press Value After Rehydration, lb*	Shedding Time, min
500	2.8	277	4
1,000	2.7	270	4
1,500	2.7	307	4
2,000	2.8	282	4½
2,500	2.9	330	5

*No significant difference at 5% level.

TABLE 2.3: TEXTURE AND REHYDRATION RATIO OF DEHYDRATED ONIONS AS AFFECTED BY COMPRESSION

Compression Pressure, psi	Rehydration Ratio*	Shear Press Value After Rehydration, lb*	Shedding Time, min
500	4.8	175	30
1,000	4.7	186	40
1,500	4.9	174	60
2,000	4.9	172	70
2,500	4.8	141	90

*No significant difference at 5% level.

TABLE 2.4: TEXTURE AND REHYDRATION RATIO OF DEHYDRATED SPINACH AS AFFECTED BY COMPRESSION

Compression Pressure, psi	Rehydration Ratio*	Shear Press Value After Rehydration, lb*	Shedding Time, min
500	8.9	105	10
1,000	8.8	116	10
1,500	7.9	120	25
2,000	7.4	102	30
2,500	7.6	117	30

*No significant difference at 5% level.

TABLE 2.5: TEXTURE AND REHYDRATION RATIO OF DEHYDRATED CARROTS AS AFFECTED BY COMPRESSION

Compression Pressure, psi	Rehydration Ratio*	Shear Press Value After Rehydration, lb*	Shedding Time, min
500	15.2	195	½
1,000	15.8	175	½
1,500	15.4	195	¾
2,000	15.2	165	¾
2,500	14.0	182	1

*No significant difference at 5% level.

TABLE 2.6: TEXTURE AND REHYDRATION RATIO OF DEHYDRATED GREEN BEANS AS AFFECTED BY COMPRESSION

Compression Pressure, psi	Rehydration Ratio*	Shear Press Value After Rehydration, lb*	Shedding Time, min
500	10.9	141	2
1,000	11.4	114	3
1,500	11.0	142	3
2,000	11.7	180	3½
2,500	11.8	149	3½

*No significant difference at 5% level.

TABLE 2.7: EFFECT OF CONDITIONING MOISTURE AND COMPRESSION FORCE ON TEXTURE AND REHYDRATION RATIO OF COMPRESSED PEAS

Conditioning Moisture, percent	Compression Pressure, psi	Shear Press Force, lb	Rehydration Ratio	Average Sensory Scores for Texture
0	0	285	4.2	6.3
8	1,000	225	4.3	5.6
8	1,500	219	4.3	5.8
8	2,000	237	4.0	5.0
12	1,000	275	4.0	5.7
12	1,500	280	3.9	6.0
12	2,000	250	3.9	6.3

TABLE 2.8: BULK DENSITIES OF AIR DRIED AND FREEZE-DRIED VEGETABLES BEFORE AND AFTER COMPRESSION

			----- Compression Ratio -----	
Product	Freeze-Dried g/cc	Compressed g/cc	Calc from Bulk Densities	Meas by Actual Fill of Cans
Peas	0.216	0.889	4.1	4
Corn	0.192	0.810	4.2	4
Onions	0.190	1.010	5.3	5
Spinach	0.038	0.419	11.3	11
Carrots	0.036	0.530	14.7	14
Green beans	0.038	0.615	16.3	16

REFERENCES

(1) Ishler, N.I. 1965. *Methods for Controlling Fragmentation of Dried Foods During Compression.* Final Report. Contract No. DA 19-129-AMC-2 (X), Technical Report D-13. U.S. Army Natick Laboratories, Natick, Mass.

(2) Brockmann, Maxwell C. 1966. *Compression of Foods.* Activities Report 18 (2), 173-177.

(3) Mackenzie, A.P. and Luyet, B.J. 1969. *Recovery of Compressed Dehydrated Foods.* Technical Report FL-90, U.S. Army Natick Laboratories, Natick, Mass.

(4) Anderson, D.B. 1935. "The Structure of the Walls of the Higher Plants." *Bot. Rev.* 1:52.

(5) Bonner, J. 1936. "The Chemistry and Physiology of the Pectins." *Bot. Rev.* 2:475.

(6) Jolyn, M.A. and Phaff, H.J. 1947. "Recent Advances in the Chemistry of Pectic Substances." *Wallenstein Lab. Communic.* 10:39.

(7) Sterling, C. 1955. "Effect of Moisture and High Temperatures on Cell Walls in Plant Tissues." *Food Technol.* 20:474.

(8) Kertesz, Z.I. 1951. *The Pectic Substances.* Interscience Publ., Inc., New York, N.Y.

(9) Reeve, R.M. and Leinbach, L.R. 1953. "Histological Investigations of Texture in Apples. Composition and Influence of Heat on Structure." *Food Research* 18:592.

(10) Fennema, O. and Powrie, W.D. 1964. "Fundamentals of Low-Temperature Food Preservation." *Adv. Fd. Res.,* 13:219

(11) Federal Specification, JJJ-0-533b. *Dehydrated Onions.*

Reversible Compression of Freeze-Dried Fruits

Technical Report 70-52-FL written for the U.S. Army Natick Laboratories by A.R. Rahman, G.R. Taylor, G. Schafer, and D.E. Westcott, and dated February 1970 covers developmental studies of the reversible compression of blueberries and cherries. Reversible compressed foods are those which can subsequently be restored to normal appearance and texture by rehydration. The process is described in more detail in the patent by *A.R. Rahman; U.S. Patent 3,806,610; April 23, 1974; assigned to the U.S. Secretary of the Army.*

The fruits prepared by this process were designed primarily to be used for making pies in U.S. Army kitchens, and the work was initiated to determine the effect of compression on the texture and overall quality of freeze-dried fruits having high sugar content. Compression ratios were determined to establish the savings possible in costs of packaging materials, handling, storage, and transportation.

GENERAL EXPERIMENTAL PROCEDURES

Sulfiting: It has been found to be generally desirable to treat the fruit with a sulfiting solution prior to compression thereof to minimize discoloration of the fruit during subsequent processing. If fresh fruit is used, the sulfiting treatment is conveniently applied just before the freezing of the fruit preparatory to freeze-vacuum-dehydration thereof. If frozen fruit is used, as is sometimes done for convenience, the frozen fruit may be treated with the sulfiting solution either in the frozen state or partially thawed. The sulfiting solution may comprise any of several known sulfiting compounds such as sodium metabisulfite, potassium metabisulfite, sodium sulfite, potassium sulfite, calcium sulfite, sulfurous acid, and liquid sulfur dioxide. Sufficient edible sulfite or bisulfite should be applied to the fruit to obtain a dehydrated fruit product having a sulfite content of about 1,000 ± 250 parts per million by weight calculated as sulfur dioxide.

Precompression Treatment: One aspect of this process relates to the discovery that the addition of moisture or a moisture-mimetic material to freeze-vacuum-dehydrated foods prior to compression thereof is not necessary in all cases to avoid shattering of the food. A compressed food product which is shattered tends to produce a mushy product upon reconstitution. Certain fruits which in the fresh state contain more than 10% sugar by weight may be compressed to a great extent without the addition of moisture or a moisture-mimetic material after freeze-vacuum-dehydration when heated in an oven or other suitable apparatus at temperatures of from 150° to 280°F for varying periods of time.

It is preferable to heat the freeze-vacuum-dehydrated fruit at a temperature of from about 200°F to about 250°F for about one minute immediately prior to compression thereof. This combination of temperature and time is particularly suitable for a continuous process. However, the upper limit of the temperature will depend largely on the tendency of the freeze-vacuum dehydrated fruit to discolor at elevated temperatures and the time the fruit is exposed to the elevated temperature. About 280°F has been found to be the maximum temperature at which freeze-vacuum-dehydrated fruits, such as pitted cherries and blueberries, may be heated for compression without causing appreciable discoloration of the skin portions of the reconstituted fruits.

The oven temperature may be as low as about 150°F, but of course as the oven temperature is reduced, the time of exposure of the fruit to the heat must be increased in order to permit sufficient time for the heat to be conducted through to the center of the fruit before it is compressed. For example, at an oven temperature of about 150°F, heating for a period of about 5 minutes is required. If the individual fruits are not heated throughout, they will crumble when compressed. Hence, the time during which the fruit is exposed to a heated environment must be correlated with the temperature of the environment in order to produce the degree of plasticity needed in the freeze-dried fruit.

Compacting Pressure: Pressures of from about 100 to about 2,500 pounds per square inch are satisfactory for compressing the heated, freeze-vacuum-dehydrated fruits to produce an acceptable compacted product. Pressures below 100 pounds per square inch do not produce sufficiently high compression ratios or densities for practical purposes, while pressures above 2,500 pounds per square inch generally result in products which are so highly compacted that they do not rehydrate rapidly or completely enough to be acceptable.

If pressures much above 2,500 pounds per square inch are used, some of the fruits will remain flat and incompletely rehydrated even after a very long rehydration time. From the practical standpoint, it is preferred that a compression ratio of at least about 6:1 be obtained in practicing the process to justify the additional process steps required for compression. In other words, it is preferred that the freeze-dried fruit be compressed sufficiently to increase its density at least about six-fold or to at least a value of about 0.6 gram per cubic centimeter. However, compression ratios as great as about 14:1 have been obtained while producing compacted, freeze-dried cherries which rehydrate satisfactorily to produce acceptable cherry pies. When the volumes of compressed freeze-vacuum-dehydrated cherries and loose frozen cherries are compared, it is possible to obtain a compression ratio as high as about 17:1. The preferred pressure range from a practical standpoint is from about 200 psi to about 400 psi since at these pressures the compaction has almost attained the maximum possible for acceptable compacted freeze-dried cherries or blueberries and has reached a point of diminishing return for increasing increments of pressure. This will be apparent in the specific procedures which follow.

Hence, it is preferable that the compressed freeze-dried cherries or blueberries have bulk densities in the range from about 0.9 to about 1.1 grams per cubic centimeter, although fruits having bulk densities as high as about 1.4 grams per cubic centimeter have been found to rehydrate well and to produce acceptable pies from the reconstituted fruits, especially in the case of freeze-vacuum-dehydrated cherries.

The process is particularly applicable to dehydrated fruit products which have been dehydrated to moisture contents appreciably less than they normally have as fresh fruits, for example, to moisture contents below about 5.0% by weight. However, the process is effective for application to freeze-vacuum-dehydrated fruits having less than about 2.0% moisture by weight. These products are the most stable in the absence of refrigeration but are most likely to shatter if compressed while heated at temperatures below about 150°F and without the addition of moisture or moisture-mimetic material.

Rehydration: In general, it is desirable for the compressed, dehydrated fruit to have a rehydration ratio at least equal to about 70% of the rehydration ratio of the uncompressed, dehydrated fruit. The rehydration ratio is defined as the ratio of the weight of the rehydrated fruit to the weight of the dehydrated fruit and is, therefore, an indicator of the extent to which the compressed dehydrated fruit can be readily rehydrated to essentially the same form as it was in prior to dehydration.

SPECIFIC PROCEDURE FOR CHERRIES

Freeze-Drying: Individually quick frozen red tart pitted (RTP) cherries, obtained from a commercial source, were sulfited by dipping in a solution of 4.0 ounces of sodium metabisulfite in 7.0 gallons of water for one minute, then draining for two minutes, to give sulfited cherries which produced freeze-vacuum-dehydrated cherries having approximately 1,000±250 parts per million by weight of sodium metabisulfite therein calculated as sulfur dioxide. The sulfited cherries were placed in a single layer on trays and freeze-vacuum dehydrated over a period of approximately 16 hours employing a shelf temperature of approximately 125°F for supplying the heat of sublimation to the frozen cherries resting thereon.

Compression: The freeze-dried cherries having a moisture content of approximately 2.0% by weight and a bulk density of 0.10 gram per cubic centimeter were heated in the dry state in an oven preheated to a temperature of 200°F over a period of one minute. The heated cherries were then immediately compressed in a Carver hydraulic press employing pressures shown in Table 3.1 with a dwell time of about 5 seconds to produce discs approximately 3⅝ inches in diameter, about one-half inch thick, weighing approximately 3.5 ounces and having bulk density values, compression ratios, and rehydration ratios as shown in Table 3.1.

TABLE 3.1: PROPERTIES OF COMPRESSED CHERRIES

Compression Pressure (lbs. per sq. in.)	Bulk Density (gm/cc)	Compression Ratios Calculated from Bulk Densities	Rehydration Ratios
0	0.10	—	2.8
100	0.91	9.1	2.4
200	1.03	10.3	2.3
400	1.06	10.6	2.1
800	1.14	11.4	2.1
1000	1.21	12.1	2.0
1500	1.27	12.7	2.0

Rehydration: The discs of compressed freeze-dried cherries were rehydrated by boiling for 2 to 3 minutes in approximately three cups of water for each disc, then permitting them to stand for 30 minutes. The rehydrated cherries were then used to produce cherry pie fillings in accordance with a commercial recipe for making cherry pie filling from reconstituted freeze-dried cherries and the filling prepared from each disc was used in making a 9-inch diameter pie. The pies were served to an expert technological panel trained in quality testing of foods. Cherry pies were prepared and tested in a similar manner using cherry pie filling prepared from uncompressed freeze-dried cherries from the same batch of cherries as that from which the compressed freeze-dried cherries were prepared.

Results: The results of the quality testing of the cherry pies prepared from the compressed freeze-dried cherries and the uncompressed freeze-dried cherries are presented in Table 3.2, the ratings being based on the so-called "hedonic" scale wherein a rating is given from 1 to 9, a rating of 1 representing "dislike extremely" and a rating of 9 representing "like extremely," and ratings in between representing various gradations between these two extremes, a rating of 5.0 being generally considered as the borderline of acceptability.

From the results, it is apparent that the cherry pies prepared from the compressed freeze-dried cherries scored almost as well as the uncompressed cherries in most instances and even better in certain respects and that in all respects they were found to be quite acceptable, i.e., having hedonic scale ratings above 5.

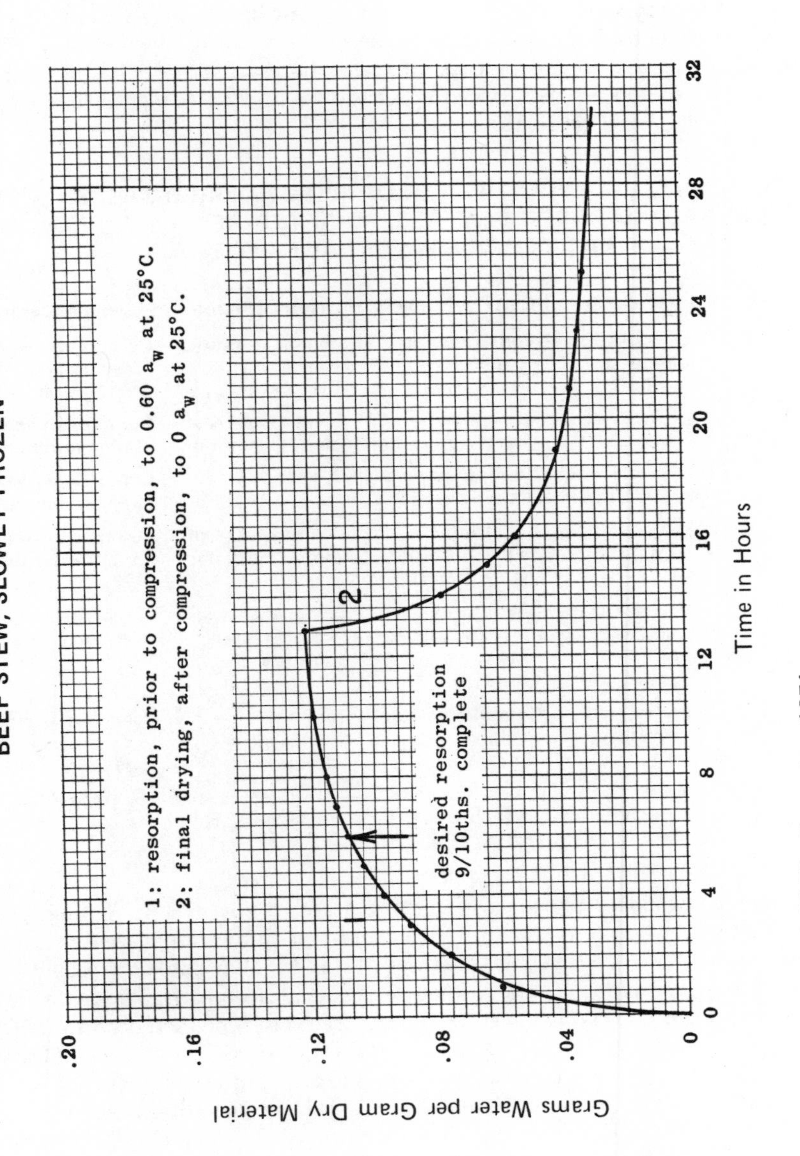

CONCLUSIONS

This study has developed a number of prototype bars which are of such quality as to warrant further study. Formulations should be studied in more detail so that the best bars for the purpose can be obtained.

For example, fat is needed in the meat products to improve flavor, increase calorie content, and decrease the sensation of dryness when the bars are eaten without rehydration. However, fat acts as a water repellant, especially when it has been dislocated by compression and smears the individual particles in the bar. While compression parameters have been investigated, more detailed studies must be conducted to elucidate interactions between all the variables in the formulations.

Obtaining the correct moisture content in the freeze-dried bars so that they are properly plasticized for compression is a problem. The most satisfactory method with most bars, particularly the meat combinations, has been to spray the product with the required weight of water and allow it to equilibrate under vacuum for three days. This method is not as precise as could be desired and requires time in production.

Compression ratios have not been determined accurately for the products. However, with the meat combination products, the ratio is approximately 3.5 to 4.5 to 1.

Nonreversibly Compressed Dehydrated Bars

Compacted, dehydrated food bars have been utilized by the Armed Forces in field rations, by astronauts during space explorations, by earth explorers, hikers and others who must carry their food supplies along with them. An outstanding advantage of rations in this form is that they provide highly concentrated nutritional values in compact and convenient forms. They also may be stored for considerable periods of time without spoilage, especially when the moisture content is sufficiently low to prevent growth of microorganisms in the compacted food bars.

One of the outstanding problems encountered with fruit bars is that when they are compacted by application of pressure in the formation of the bars, if the moisture content is as low as is desirable for stability, the fruit bars become so hard that they are extremely difficult to eat directly without rehydration. In some cases, they have been known to become so hard that when an attempt is made to eat them directly, they have caused breakage of teeth.

On the other hand, if enough moisture is left in the compacted fruit bars to permit direct eating of the bars without the danger of damage to teeth, the fruit bars may be unstable in long-term storage. Further, they may be very difficult to form because of the extrusion of the fruit pulp from the mold during compression in forming the bars.

Technical Report 71-60-FL, prepared for the U.S. Army Natick Food Laboratory by A.R. Rahman, G. Schafer, P. Press and D.E. Westcott, entitled *Nonreversible Compression of Intermediate Moisture Fruit Bars*, July 1971, covers developmental studies of nonreversible compressed fruit products including dates, figs, raisins, nuts, and combinations of these fruits.

TECHNICAL REPORT

Materials

Food ingredients used during the course of these studies such as dates, figs, maraschino cherries, golden raisins, sesame and nuts were locally purchased. During preliminary work, products such as apricots, prunes and brown raisins were also studied, but were found to discolor excessively during storage at 100°F. The dates, figs and maraschino cherries were chopped into pieces of approximately ¼ inch. All fruits were then dehydrated to a practical range suitable for compression. This step was necessary since it was impractical to compress fruits with original moisture ranging between 15 and 35% due to excessive extrusion of pulp.

Successful compression of intermediate moisture fruits (15 to 30% moisture) was accomplished when the moisture content was reduced by dehydration to 7 to 14%. The bars were hard and difficult to chew when the moisture content was reduced below 7%. All fruits used in these studies were dehydrated to approximatly 8% before compression. The bars were formulated as shown in Table 6.1.

Compression

43 grams of thoroughly mixed ingredients were compressed into 1" x 3" x ½" bars with a hydraulic laboratory press (Carver) using a compression force of approximately 200 pounds per square inch. The molds were sprayed with a food grade lubricant (lecithin) in order to reduce sticking and to facilitate removal of the bars.

After compression the bars were sealed in a flexible pouch (polyester, 0.5 mil; aluminum foil, 0.35 mil; polyolefin, 3.00 mil) at a vacuum of approximately 27 inches of mercury. Initially, fourteen different fruit bars were formulated. Preliminary testing indicated that the bars were difficult to bite and some fragmented readily, especially the date bars. Lecithin at a 2% level was found to improve the texture significantly.

Bulk Density

Bulk density was measured by dividing the weight of the loose (or compressed) product by its respective volume to yield grams per cubic centimeter. Compression ratio was determined by dividing the volume of the uncompressed product by its compressed volume.

Sensory Evaluation

Sensory evaluations of product quality were conducted by ten trained technologists. During each test session, the panel members were given two samples, one-half bar each, in a balanced random order. Since the bars were to be

TABLE 6.1: FRUIT BAR FORMULATIONS

Date Cherry Bar	
Dates	49%
Maraschino cherries	49%
Lecithin	2%

Date Fig Almond Bar	
Dates	39%
Maraschino cherries	25%
Figs	24%
Almonds	10%
Lecithin	2%

Date Sesame Bar	
Dates	78%
Blanched sesame seed	20%
Lecithin	2%

Raisin Bar	
Raisins	98%
Lecithin	2%

Date Cherry Orange Bar	
Dates	53%
Maraschino cherries	30%
Orange peel	15%
Lecithin	2%

Date Fig Bar	
Dates	49%
Figs	49%
Lecithin	2%

Date Sesame Bar	
Dates	78%
Sesame	20%
Lecithin	2%

Fig Cherry Pear Orange Bar	
Figs	68%
Cherries	15%
Pears	10%
Orange peel	5%
Lecithin	2%

Raisin Bar	
Golden raisins	98%
Lecithin	2%

judged independently, attempts were made to always serve nonrelated items. Sensory panel evaluation for color, flavor and texture were conducted using the trained panel and a 9 point quality rating scale (1 = extremely poor and 9 = excellent).

Since the United States Marine Corps requested the development of four menus for an Emergency Food Packet, only four fruit bars were selected (Table 6.2) so that one bar could be included in each menu. Each of the four, flexibly packaged bars was packed together with other food components in a second pouch and hermetically sealed to form a packet. The packets were stored at 100°F for up to 12 months or at 70°F for up to 24 months. No direct comparisons between the foods stored at 70°F and those stored at 100°F were made. However, standards of the fruit bars (stored at 40°F) were presented to the panel at each session for comparison. The study was designed so that on each day of testing all four menus were evaluated. Each panel member received the components of a single menu in a balanced random order, testing all four different menus in a four day period.

Texture

The Instron Universal Testing Apparatus, Floor Model TT-DM with a 500 kg cell was used. The samples were penetrated to 50% of initial thickness at a speed of 2 cm/min using a cylindrical 0.75 cm punch. Results were expressed as (1) firmness, which is the force at 50% penetration, (2) toughness, which is the work expended for a 50% penetration, and (3) maximum force in kg, which is used as an empirical measurement of hardness.

The caloric value was determined by the Parr Oxygen Bomb Calorimeter. Moisture content was determined in accordance with the AOAC method using a vacuum oven. The results were statistically analyzed and the least significant difference was determined as applicable.

Results

During these studies all fruits were dehydrated to a moisture content of approximately 8% before compression. However, preliminary work indicated that fruits such as dates, figs and raisins can be successfully compressed when the moisture content ranged from 7 to 14%.

Results of the technological panel evaluations indicate that there is no significant difference in color, flavor and texture of nine different fruit bars before and after storage for 6 months at 100°F. All ratings ranged between 5 and 7 (fair to good); however, most of the ratings ranged between 6 and 7. The four fruit bars selected for further quality evaluations as components of the Marine Corps Food Packet were date cherry bar, date fig almond cherry bar, raisin bar and date sesame bar. As shown in Table 6.2, the color, flavor and texture of date cherry, date fig, date sesame and raisin bars stored for 12 months at 70°F, as determined by a technological panel, did not change. However, after storage

TABLE 6.2: AVERAGE RATINGS (TECHNOLOGICAL PANEL) OF QUALITY OF FRUIT BARS AS AFFECTED BY STORAGE AT 70°F

Item	Conditions	Color	Flavor	Texture
Date Cherry	initial	7.3	7.1	7.0
	12 mo.	6.6	6.8	6.7
	24 mo.	6.6	6.2	6.1
	LSD	NS	NS	0.6
Date Fig Cherry Almond	initial	7.2	7.2	7.0
	12 mo.	7.0	6.9	6.7
	24 mo.	6.9	6.6	6.6
	LSD	NS	NS	NS
Date Sesame	initial	7.2	7.0	7.1
	12 mo.	6.8	6.9	7.1
	24 mo.	6.6	6.2	6.3
	LSD	NS	0.6	0.6
Raisin	initial	7.1	7.1	7.0
	12 mo.	6.6	6.8	6.9
	24 mo.	6.4	6.4	6.4
	LSD	NS	0.4	0.4

TABLE 6.3: BULK DENSITY AND CALORIC VALUE OF FRUIT BARS

Fruit Bar	Uncompressed		Compressed			
	Bulk density gm/cc	Calories per cc	Bulk density gm/cc	Calories per cc	Compres-sion Ratio	Calories per gram
Date Cherry	0.43	1.6	1.32	5.0	3.0	3.8
Date Fig	0.49	2.1	1.26	5.5	2.6	4.4
Date Sesame	0.62	2.8	1.27	5.8	2.0	4.6
Raisins	0.57	2.0	1.54	5.7	2.7	3.7

of 24 months, the flavor of the date sesame bars and the raisin bars was rated significantly lower than initially, although the ratings were above 6, which is considered a good quality product. In addition, the texture of the date cherry, date sesame and raisin bars was significantly lower than initially, although still above 6. No significant change was exhibited in the color of the bars throughout the storage period.

When the packaged fruit bars were combined with other food components in a secondary packet to form a single meal and stored at 100°F for 6 months, the color and flavor of the raisin bars exhibited significantly lower ratings than initially. The rest of the bars were unchanged.

After 12 months of storage, the color and flavor of date cherry, date fig and raisin bars received significantly lower scores, whereas the date sesame showed a significantly lower score for color only. Texture, except for the raisin bar, was not affected.

Results of subjective as well as objective tests on the texture of date bars indicate that the addition of lecithin significantly improves the texture of the bar. Therefore, the addition of lecithin was incorporated in the formulations of all bars during the course of these studies.

The compression ratios of the fruit bars ranged from 2:1 to 3:1 (Table 6.3). This is a considerable reduction in volume which ultimately results in savings in packaging materials, storage space and perhaps shipping costs. The caloric value of the four bars ranged between 3.7 and 4.6 calories per gram. The uncompressed products ranged between 1.6 and 2.8 calories per cubic centimeter, whereas in the compressed product this range was significantly increased to between 5.0 and 5.8 calories per cubic centimeter (Table 6.3).

ILLUSTRATIVE FRUIT BARS

A.R. Rahman and G.R. Schafer; U.S. Patent 3,705,814; December 12, 1972; assigned to the U.S. Secretary of the Army describes the process for making several of these fruit bars in more detail. Their directly edible, compacted and dehydrated fruit bar is comprised of one or a plurality of fruits suitably subdivided and coated with a lecithin or modified lecithin-containing composition and dried to a moisture content in the range of from 7 to 14% on a weight basis, then compressed at a pressure of from 200 to 3,000 psi into bar form of such dimensions as to facilitate the direct eating of the bar without prior rehydration thereof.

A relatively small quantity of lecithin or a modified lecithin, usually less than 5% by weight, when well distributed through a fruit bar on the surfaces of the subdivided particles that are compressed together to form the fruit bar, produces a change in the texture of the fruit bar such that the resulting compacted fruit bar can be easily bitten through and chewed without damage to the teeth of the consumer and with enjoyment in contrast to the difficulty with which a similar fruit bar lacking the lecithin or modified lecithin can be eaten directly. The lecithin or modified lecithin used in making the compacted, dehydrated fruit bar is an edible lecithin or modified lecithin. It may be derived from various natural sources, but the lecithins or modified lecithins produced from the natural lecithin of soybeans have been found to be particularly effective for this purpose.

Generally speaking, the fruit is subdivided to form particles of from about ⅛ to ½ inch in the longest dimension and the particles are dried to from 7 to 14% moisture content by weight prior to the application of the lecithin or modified lecithin-containing composition. A sufficient amount of the solution or suspension of the lecithin or modified lecithin is applied to the particles of fruit to obtain a reasonably uniform coating on the particles of from 1 to 5% of lecithin or modified lecithin on a weight basis. About 2% lecithin or modified lecithin has been found to be particularly effective.

The pressure required in compressing the subdivided fruit coated with lecithin or modified lecithin will, in general, depend on the amount of moisture in the fruit. The lower the moisture content of the fruit down to as low as 7%, the higher the pressure required, up to 3,000 psi, to obtain a fruit bar of the proper degree of adhesion. The higher the moisture content of the fruit, up to as high as 14%, the lower the pressure required, down to 200 psi, to obtain a fruit bar of the proper degree of adhesion. The type of fruit has some effect on the pressure required, figs and pears in general requiring higher pressures than most other fruits.

The process has been found to be particularly effective with dried dates, raisins, cherries, figs and pears. However, it may also be applied to other fruits; also other components, such as edible seeds and nuts, may be added to the fruits in order to impart variety and interest to compressed fruit bars. In addition, cereals, proteins, fats, chocolate, spices and various other flavoring or chemical additives may be incorporated in the compressed fruit bars in minor proportions in relation to the fruit components.

Example 1: Dates were diced to form particles having a maximum dimension of 0.5 inch. The particles of diced dates were air dried to a moisture content of 8%, then spread out in a pan and spray-coated as uniformly as possible over the surfaces of the particles with PAM (an aerosol solution of lecithin) until an add-on of about 2.0% by weight was obtained. The particles coated with the lecithin solution were placed in a mold and compressed at 400 psi pressure to form a bar of the dimensions of 3 x 1 x 0.5 inches.

The bars were stored overnight in a closed airtight container, then tested with an Instron Universal Testing Apparatus, Floor Model TT-DM, using a 500 kg cell. The bars were penetrated at 50% of their initial thickness at a speed of 2 cm per minute using a cylindrical, flat-surfaced, punch having a diameter of 0.75 cm. The results in Table 6.4 were obtained. Firmness is represented by the force at 50% penetration. Toughness is represented by the work expended in penetrating to 50% of initial thickness. Hardness is represented by the maximum force applied during the penetration.

TABLE 6.4

	Firmness (force at 50 percent penetration), kg.	Toughness (work expended for 50 percent penetration), kg·cm.	Hardness (maximum force), kg.
Sample with lecithin:			
Average	5.1	3.86	5.6
Sample without lecithin:			
Average	6.9	5.09	7.4

Example 2: Dates, figs, cherries and almonds were diced to form particles having a maximum dimension of 0.5 inch. These particles as well as sesame seeds were mixed in the proportions shown in Table 6.5, and the separate mixtures were spread out in pans and spray-coated in substantially the same manner as for the date particles in Example 1. The separate mixtures were placed in molds and compressed at 1,000 to 1,500 psi to form bars measuring 3 x 1 x 0.5 inch.

These bars were subjected to technological taste panel testing initially and after periods of storage at 100°F as shown in Table 6.5, with the results shown in terms of a 9-point hedonic scale. The numbers in the table represent the average of ten taste testers' evaluations of the samples, with 1 representing extreme dislike, 9 representing extreme like, and a value of 5 representing borderline acceptability.

TABLE 6.5

Sample	Ingredient	Percent	Storage	Panel ratings of— Color	Odor	Flavor	Texture
A	Date	39	[1]0	7.3	7.3	7.4	7.1
	Fig	24	1½	7.5	7.1	7.0	7.0
	Cherry	25	3	7.1	6.8	7.0	6.8
	Almond	10	6	6.9	6.6	6.9	6.4
	Lecithin	2					
B	Date	49	[1]0	7.1	6.3	6.6	6.8
	Cherry	49	1½	7.3	7.1	7.3	6.3
	Lecithin	2	3	7.5	7.3	7.2	6.7
			6	7.3	7.3	7.2	6.7
C	Date	78	[1]0	7.4	6.7	6.9	7.0
	Sesame	20	1½	7.1	6.5	6.2	6.9
	Lecithin	2	3	7.4	6.7	6.2	6.8
			6	6.8	6.6	5.5	6.8

[1] Initial.

Some of the mixtures of particles were compressed under similar pressure conditions but without application of lecithin thereto. The resulting bars were so hard to bite into that no satisfactory taste test comparison could be made either initially or after the storage periods applied to the bars containing lecithin since to carry out such tests would have gravely endangered the teeth of the members of the panel to breakage.

ILLUSTRATIVE CHEESE-BASED BAR

It has long been recognized that cheese is a valuable component of the human diet, in that it is rich in fat and proteins and highly taste acceptable to almost every civilized human being. However, cheese has the following drawbacks: it has a tendency of staling and/or becoming moldy after a relatively short shelf life, which totally deprives it of its taste acceptability. Moreover, the low carbohydrate content of cheese is wholly insufficient to sustain human life. Consequently, cheese is usually consumed together with a carbohydrate-rich food component, such as bread, as in the conventional cheese sandwich. However, this combination of bread and cheese is unsuitable for a survival ration, as bread will stale even more quickly than cheese.

L. Jokay; U.S. Patent 3,121,014; February 11, 1964; assigned to the U.S. Secretary of the Army has developed a dehydrated, carbohydrate-enriched, palatable, cheese-based food bar which is shape-retaining, can be consumed without rehydration, and is suitable as a survival or other ration for military purposes, as well as for such civilian users as explorers, sailors, hunders, campers and the like.

The cheese-base dehydrated food bar is prepared from the following ingredients.

(1) A fatty cheese. Blue cheese, a popular taste cheese is preferred. Other fatty cheeses with well-aged flavor, e.g., of the Roquefort, Swiss or aged American type may also be used; they contain more than 25% fat and are thus included in the term "fatty cheese."

(2) Dehydrated ground potatoes. Preferably, the dehydrated potatoes for use in the food bar are produced from cooked, diced potatoes that were subjected to a sulfiting and calcium chloride pretreatment before dehydration (to prevent discoloration), and ground after dehydration to a fineness of about U.S. Standard No. 8 to No. 30 sieve size.

(3) Dehydrated water-soluble starch; preferably of the precooked, gelatinized potato starch type.

(4) Fat, e.g., shortening, preferably hydrogenated. Vegetable fats (or mixtures thereof) may be used for shortening in whole or in part.

The foregoing ingredients are compounded in such a way that the final composition has the following nutrient contents:

	Parts by Weight (Approximate)
Proteins	7-9
Fats	13-15
Carbohydrates	70-75

Cheese being the principal protein source for the food bar, it may be necessary to add a small additional amount of soft cheese, such as uncreamed cottage cheese to the preferred formulation in order to raise the protein content to the foregoing desired standards.

Example: A food bar containing the foregoing ingredients is preferably compounded as follows, using a food mixer. About 85 parts of blue cheese are dispersed in an excess of water (about 400 parts) above room temperature (preferably 140°F), and about 200 parts of dehydrated procooked ground potatoes are then mixed into this slurry until a smooth paste is produced.

A second batch is prepared by mixing about 30 parts of hydrogenated shortening with an excess of water (about 200 parts), also above room temperature sufficient to melt the shortening (preferably about 140°F), and about 112 parts by weight of potato starch are blended with this mixture of water and melted shortening, which causes the starch to rehydrate.

This second batch is then combined with the first batch and blended above room temperature (preferably 140°F) until a homogeneous mixture is obtained. This mixture is then chilled to a solid pliable mass, without freezing. This mass is molded into bars of the desired size, preferably by pressure molding on a hydraulic press. The size of these bars should be larger than the desired size of the final dehydrated product, as they will shrink during the following dehydration step. Thus, if a dehydrated bar of an area of 90 x 46 mm is desired, the press mold should have an internal dimension of 95 x 51 mm. Similarly, the thickness of the molded (undehydrated) bar should be 5 to 10% in excess of the desired final thickness; thus, if the final thickness is intended to be 14 mm, the bars should be initially molded to a thickness of 15 to 15.5 mm.

The molded bars are then freeze-vacuum dehydrated by following techniques summarized below. The bars are frozen to -15° to -20°F and vacuum-dehydrated in a chamber equipped for obtaining and holding a very low absolute pressure (high vacuum). The chamber should be equipped with shelves which can be heated or cooled by means of a circulating liquid medium. The drying chamber should be thoroughly cleaned and free of all traces of foreign odors coming from a disinfectant, a washing solution, or from products previously dried. The shelves should preferably be chilled, using ice water or other refrigerant before and during loading and while the vacuum is being drawn on the oven. However, if cooling of the shelves is impossible, the frozen trays should be kept frozen and as cold as possible before loading the oven. To avoid thawing of any portion of the frozen cheese bars while they are being transferred to the dryer, arrangements should be made to accomplish the transfer quickly and to draw a vacuum of 1.5 mm or lower, not to exceed about 8 minutes after completion of the transfer.

About 30 minutes after reaching 1.5 mm absolute pressure or less (preferably 600 to 800 microns), heat is applied to the circulating medium. The final temperature of the circulating medium should be 110°F. This temperature should be reached in about 4 hours or more. The rate of heating should be so regulated that the plate temperature for the first 2 hours should be about 50°F and about 90°F for the next 2 hour period. The absolute pressure in the drying chamber at no time should exceed 1,500 microns (preferably 750 microns). When these conditions have been established, they are to be maintained throughout the drying process, or until the product contains not more than about 4% moisture, preferably 2 to 2.5%.

A typical bar prepared as above contains about 7 parts protein, 14 parts fat, 72 parts carbohydrate, 4 parts ash, 1 part salt, and 2 parts moisture per 100 parts by weight.

The bars, when properly wrapped and sealed (preferably heat-sealed under vacuum conditions or in an atmosphere of low oxygen content) in conventional wrapping material suitable for cheese foods, such as polyethylene terephthalate sheeting or pouches, will have a shelf life of many months without losing their taste acceptability.

Miscellaneous Bars

ALL-PURPOSE MATRIX

P.J. Mech, R.W. Groncki and T.A. Smith; U.S. Patent 3,336,139; August 15, 1967; assigned to Evans Research and Development Corporation have developed an all-purpose matrix which permits acceptable food bars to be made of a large variety of menu items such as the following.

Main Dishes: Beef stew, chili con carne, chicken and rice, shrimp creole, Welsh rarebit, chicken a la king, etc.
Soups: Cream of mushroom soup, beef barley soup, vegetable soup, clam chowder, etc.
Puddings: Tapioca pudding, chocolate pudding, plum pudding, banana cream pudding, etc.
Beverages: Coffee with cream and sugar, orange juice, apricot nectar, etc.

The matrix contains a combination of a polymer and a sugar. The polymer and sugar are combined in a water solution, dried, blended with the desired food components and then molded together into the food bar.

Not all polymers are capable of being used in this matrix, but only those that are water-dispersible, film-forming, edible polymers, such as: methylcellulose, polyvinyl alcohol, polyvinyl pyrrolidone, hydroxyethylcellulose, and ethers of methylcellulose and hydroxypropylmethylcellulose, and mixtures thereof. Sodium carboxymethylcellulose is particularly preferred because of its lack of taste.

The sugars that have been found to be effective are those having relatively low flavor levels such as sucrose, dextrose, fructose and lactose; lactose is particularly efficacious because of its lower sweetness level. The ratio of the matrix may be from 85 parts of sugar:15 parts of polymer to 99.5 parts of sugar:0.5 part of polymer. The parts are by weight. 99:1 is a particularly good ratio for the best universal result.

The sugar and polymer each possess some adhesive properties, but the combination of the two components, as specified herein, produces unusual and unexpected binding effects, e.g., stronger, more plastic, more compatible and adaptable than the added effect of the two components if used separately.

The all-purpose matrix is prepared by: (a) dissolving the sugar in water, (b) dissolving the polymer in water, (c) mixing (a) and (b), and (d) drying the solution. In lieu of dissolving the sugar and polymer separately and subsequently mixing the two solutions, they may be dissolved in the same solution of water. It is necessary that enough water be present so that the sugar is dissolved and the polymer well dispersed. For example, 99 g of sugar and 1 g of polymer may be made into 1 liter of solution. More or less water may be used, but the larger the quantity of water, the longer the drying time of the solution.

The solution containing the sugar and polymer may be dried by any desired method, such as freeze-dried, tray-dried or spray-dried. The nutritional components are then combined with the dried matrix. The food components may be precooked or raw, in a dried or natural state, triturated or sliced or minced, or in whole form. The limiting factor is the moldability of the bar. If the food component is too moist, the matrix may be unable to absorb all the moisture and maintain the bar form; a food, too bulky in shape, may also affect the capacity of the bar to remain in such form. However, the more surface area of the food available for contact with the matrix, the better the physical characteristics of the resulting bar. This result was also found to be true with regard to food components having lower moisture content, e.g., dehydrated foods.

The matrix and food components are mixed by any mechanical means, e.g., stirring, Hobart mixer, etc. The best results are obtained when the components are thoroughly mixed, so that a substantial degree of homogeneity is achieved. This is easier to achieve if the matrix is in the form of a free-flowing powder.

A satisfactory bar is achieved if the bar contains 10% (by weight) matrix; in a few, isolated cases this figure may be as low as 5 to 8%. The best overall results are obtained if the matrix constitutes from 13 to 20% of the bar. The quantity of food component chosen to be added to the matrix is a matter of taste, desired caloric and nutritional value, etc. It has been found that the finer the particles of the dehydrated materials, the greater the amount that may be combined with the matrix.

The matrix-component combination is then molded under pressure. Any of the usual pressure molding processes may be utilized, such as tableting machines and other commercial pressure molding machines and techniques, or manual apparatus and methods. The degree of pressure in the molding process produces bars of varying physical properties. In general, the more pressure exerted, the harder the consistency of the bars. However, too high a pressure, for example 20,000 psi, will produce a hard, easily broken, difficult-to-dissolve bar in most instances; too little a pressure, for example 500 psi, will not usually affect a bar firm enough to hold its shape and form. A pressure of 6,000 psi produces a universally satisfactory bar.

A general procedure for manufacturing a bar is as follows. A predetermined quantity of the food component is weighed out together with the matrix and the two are dry blended. In most cases a small quantity of water is added to the dry blend to activate the matrix; the water does not have to be removed. Some of the water may be replaced by glycerol to aid in rehydration. A predetermined amount of the food component-matrix combination is weighed out and then pressure molded. The dried matrix is added and blended thoroughly with the food component, for example using a Hobart-type mixer at low speed. To facilitate mixing and to activate the binding properties of the matrix, glycerin is added. The following example will illustrate a method of preparation of this all-purpose matrix.

Example 1: 99 g of lactose were dissolved in water, 1 g of sodium carboxyethylcellulose was dispersed in water. The two solutions were combined, blended and diluted with water to 1 liter of solution. The liter of solution was then tray-dried and the resultant was ground into a free-flowing powder, thereby producing an all-purpose matrix. The above formed matrix is combined with various components to form diverse food bars as shown in the following examples.

Example 2: Canned beef stew is freeze-dried and granulated. 1,000 g of the granulated, freeze-dried stew is combined and blended with 140 g of matrix and 80 g of water, by mechanical mixing. The blended material is then compression molded at 6,000 psi to form a beef stew food bar.

Example 3: Chocolate pudding powder is reconstituted, freeze-dried and granulated. 1,000 g of the granulated pudding is combined and blended with 140 g of matrix and 120 g of glycerin by mechanical mixing until it is homogeneous. It is then compression molded at 6,000 psi to form a chocolate pudding food bar.

Example 4: 1,000 g of beef barley dry soup mix are combined and blended with 140 g of matrix and 80 g of glycerin by mechanical mixing. The blend is then compression molded at 6,000 psi to form a beef barley soup food bar.

Example 5: 1,000 g of freeze-dried tomato juice is combined and blended with 100 g of starch, 140 g of matrix and 60 g of glycerin by mechanical mixing. The blend is then compression molded at 6,000 psi to form a tomato juice food bar.

COMPRESSED, DEHYDRATED BREAD

J.J. Mancuso, A.G. Bonagura and R.M. Sorge; U.S. Patent 3,512,991; May 19, 1970; assigned to General Foods Corporation have developed a process in which compressed, dehydrated bread is prepared by conditioning the bread at a reduced temperature, compressing the bread to about one-half to one-third its original volume, freezing the compressed bread and freeze-drying the frozen, compressed bread. The resulting product retains its compressed state and rehydrates rapidly, reverting to its original size and state of freshness.

In order that the process may be better understood, the following examples will serve to illustrate specific features of it. In the examples where bread was compressed and dehydrated, the crust was first removed to facilitate the rehydration step. The bread which was used in the examples was bought on the open market and was already baked and sliced. The cake and waffles were prepared from commercial dry mixes.

Example 1: Bread slices which had the crust removed, having a thickness of 12 mm, were autoclaved in a Castle autoclave at 10 psig for ten minutes. The autoclaved slices were then wrapped in aluminum foil and conditioned overnight at 35°F. After conditioning, the bread slices were removed from the aluminum foil and were compressed to 4 mm in thickness by means of platens cooled with Dry Ice to -28°F and frozen. The bread slices were freeze-dried to a final moisture content of approximately 10%. The bread slices were rehydrated by being dipped in water for 6 seconds. A slightly moist slice of bread was obtained. The normal crumb texture and appearance was retained.

Example 2: The procedure described in Example 1 was repeated except that the conditioning took place at 0°F, 15°F and 70°F. These samples were dipped in water for 6 seconds. The following results were obtained.

Conditioning Temperature	Product Condition
0°F	Fully hydrated, moist crumb
15°F	Fully hydrated, slightly moist crumb
70°F	Not fully hydrated crumb

This example illustrates that bread which was conditioned at 0°F and also at 15°F hydrated satisfactorily, while bread which was conditioned at 70°F did not give a satisfactory product.

Example 3: Sponge cake slices having a thickness of 12 mm were autoclaved in a Castle autoclave at 10 psig for ten minutes. The autoclaved slices were then wrapped in aluminum foil and conditioned overnight at 25°F. After conditioning, the cake slices were removed from the aluminum foil and were compressed to 4 mm by means of platens cooled with Dry Ice to -28°F and frozen. While still in the frozen condition, the cake slices were freeze-dried to a final moisture content of approximately 7%. The cake slices were rehydrated by being dipped in water for 3 seconds. A slightly moist slice of sponge cake was obtained. The normal crumb texture and appearance was retained.

Example 4: Waffles having a maximum thickness of approximately 15 mm were wrapped in aluminum foil and conditioned overnight at 25°F. After conditioning, the waffles were removed from the aluminum foil and were compressed to 5 mm by means of platens cooled by Dry Ice to -28°F and frozen. While still in the frozen condition, the waffles were freeze-dried to a final moisture content of approximately 10%. The waffles were rehydrated by being dipped in water for 6 seconds. A slightly moist waffle was obtained. Normal texture and appearance was retained.

COATED, DEHYDRATED, COMPRESSED FOODS

N.E. Harris; U.S. Patent 3,726,693; April 10, 1973; assigned to the U.S. Secretary of the Army has discovered that the structure of various dehydrated food items which are brittle and fragile and tend to crumble or disintegrate can be stabilized or strengthened by means of an edible, nutritious coating. The coating can be easily applied to the food items and will not cause any increase in toughness or hardness of the food item provided that the moisture level of the food item is maintained below a critical level. The preferred coating on certain confectionary and farinaceous food products is an aqueous emulsion containing edible oil, sodium caseinate, glycerin and gelatin.

Food blocks or cubes compressed from powdery or granular food materials and having a low moisture content are also prone to crumble or shatter. These and other dried products can be stabilized as to structure in accordance with this process with a coating in such a manner as to preclude subsequent hardening of the product.

Example 1: A suitable coating to stabilize the structure of low moisture or dried compressed food products is formulated as follows.

	Percent by Weight
Edible vegetable oil (400 hr AOM stability)	9.7
Sodium caseinate	9.7
Glycerin	2.8
Gelatin (275 bloom)	2.2
Water	75.6

The coating emulsion is prepared by adding one-half of the sodium caseinate to the oil which has been heated to a temperature within the range of 150° to 160°F. The remainder of the caseinate is added to the oil and mixed until all the dry particles are coated with oil. The gelatin is soaked in cold water and allowed to swell, and is thereafter heated to 140° to 150°F. Glycerin is added to the heated gelatin solution and mixed. About one-fifth of the gelatin-glycerin water solution is added to the oil-sodium caseinate mixture and blended, and the remainder of the solution is added and blended at moderate high speed. Entrapped air is removed by heating the mixture in a water bath to a temperature of 190°F and slowly stirring for 5 minutes. The coating emulsion is maintained at a temperature of from 160° to 180°F for coating application. The emulsion is stable for several months if held at a temperature of 40°F.

Example 2: A cocoa flavored coating which may be used in this process is formulated as follows.

	Percent by Weight
Edible vegetable oil (400 hr AOM stability)	35.0
Sodium caseinate	12.0
Glycerin	7.0
Gelatin (275 bloom)	6.0
Sucrose	33.7
Cocoa	5.5
Vanilla	0.5
Citric acid	0.1
Parabens (3 parts methyl and 1 part propyl)	0.1
Potassium sorbate	0.1
Water, 65 ml/100 g of coating formula	

The cocoa coating is prepared by heating the oil to 150°F and mixing the heated oil with sodium caseinate, sucrose and cocoa. The gelatin is allowed to swell in cold water and then heated to 150°F. Glycerin, vanilla, parabens and potassium sorbate are added to the oil slurry and mixed until a stable emulsion forms. The citric acid in 1 ml of water is added to the emulsion and mixed, and a vacuum is drawn on the coating emulsion to eliminate any entrapped air. The emulsion is heated to and maintained at a temperature of 160° to 180°F for coating application.

Example 3: This example describes the preparation of a compressed, dehydrated, imitation ice cream mix cube. An ice cream mix base is formulated from the following ingredients.

	Percent by Weight
Coconut oil	12
Nonfat dry milk	11.2
Sugar	15.0
Gelatin (275 bloom)	0.3
Emulsifier	0.2
Water	61.3

The ice cream mix base was prepared by mixing all the ingredients thoroughly, heating to 165°F for 30 minutes, homogenizing at 2,000 psi in the first stage and 500 psi in the second stage, and cooling to 40°F. A vanilla flavored mix is prepared by mixing 7.0 g of ethyl vanillin to 24 lb of cooled base mix. The flavored mix is blast frozen and then freeze-dehydrated at a maximum pressure of 1.5 mm of mercury and a maximum temperature of the dry product of 120°F.

The resulting dried product is ground to yield a uniform free-flowing powder. The powder is then compressed into ¾ inch cubes weighing 5 g and coated with the emulsion coating of Example 1. The cubes are dipped into the heated (160° to 180°F) emulsion, drained, frozen and freeze-dehydrated in the manner described for the flavored mix until the moisture content was 1.52%. Similar coated cubes were dehydrated to 2.55% moisture.

Cubes were then packaged, six to a package, under 29 inches vacuum in heat sealed 1 mil Mylar/2 mil polyethylene laminate flexible pouches. The high moisture and low moisture cubes were packaged separately. The packaged samples were subjected to a temperature of 140°F for three hours to determine storage stability. The cubes having a moisture content of 2.55% upon examination were found to have hardened significantly, so hard in fact that they were difficult to bite or chew without extreme jaw pressure. The low moisture cubes (1.52%) were found not to have changed in quality and did not undergo any hardening as compared with a noncoated control.

Example 4: Compressed cereal bars prepared in accordance with Military Specification MIL-C-3483B and consisting principally of corn flakes and puffed rice (class 8 of the specification) were cut into cubes approximately ¾ inch in size. The cubes are dipped into the coating material of Example 2, which is heated to a temperature of 170°F in a water bath. The coated cubes are freeze-dried as in Example 3 with some of the cubes being dried to a moisture level of 2.5% and others being dried to a moisture level of 1.5%.

The cubes are packaged in cans which are flushed with nitrogen to reduce the oxygen content of the headspace gas to not more than 2%, and then sealed. The canned samples were stored for 3 months at 85°F. At the end of this time, the 2.5% moisture cubes had hardened significantly during storage and were practially inedible because of the rock-hard quality of the cubes. The low moisture (1.5%) cubes were found to be virtually unchanged and did not harden as did the higher moisture cubes.

COMPRESSED, BAKED, HIGH-PROTEIN FOOD

H. Corey, A. Bakal, K. Konigsbacher and D. Schoenholz; U.S. Patent 3,812,268; May 21, 1974; assigned to Foster D. Snell, Inc. have developed a process to provide compounded food products which can be compressed to a small fraction of their original volume and which on admixture with water or other liquid will swell to a volume, size and structure resembling the original. They were particularly interested in products containing a high percentage of protein and low percentages of fat and carbohydrate.

It was found that compounded food products, especially suited for compression, may be produced from a dry mix of ingredients comprised of at least 18% gluten or other vegetable protein and a relatively high percentage of a polyhydric alcohol (at least 10%), the percentages being calculated on the total weight of the product. When this mixture is mixed with water to produce a doughy mass which is baked and dried, an open cellular product is obtained having a density of 12 lb/ft^3 or more, depending on the relative ratio of the ingredients and the processing conditions employed. Usually such products have densities of 10 to 40 lb/ft^3.

Such products after being compressed to a small fraction of their original volume will recover their original volume, geometry and structure upon being immersed in water or other suitable liquid, or upon being moistened with any of these liquids and then heated in an oven at a suitable temperature above ambient, or, if the formulations contain a suitable amount of hard fat or other solid but meltable plasticizer, upon heating in an oven as above. Such food products are particularly advantageous for foods for military and space operations and for rations in any situation where small volume is important or where the expansion mechanism may be advantageous.

A typical production of such products is carried out by mixing the ingredients with appropriate amounts of water to produce a doughy mass. Depending on the desired final shape, the dough is cut into pieces or forms and baked in an oven or cooked, thus causing expansion of the system as well as a locking-in of the structure. In this manner, a sponge-type structure is obtained. The density of the product can be controlled by the baking temperature, the aeration given to the dough prior to baking and the presence of leavening agents, if any, or by controlling the pressure upon the product during baking.

Thereafter, the baked or cooked product is dried by any suitable dehydration method (air-drying, freeze-drying, vacuum air-drying, etc.) to a moisture content of 4 to 8% by weight. This level of moisture is a key factor for successful compression of the product. Too low a moisture content results in shattering and cracking during compression, while too high moisture content results in permanent deformation or even flow of the product under compression. The compression pressure (2,000 to 5,000 psig) results in a 20 to 80% reduction of the original volume. The compressed product is then locked into this compressed state by drying to a moisture content of 2% or less.

The key to the structural fidelity and the behavior of these products is associated with the relatively high vegetable protein content as exemplified by gluten, soybean protein and the like. Also, animal proteins can be incorporated in the products in lieu of gluten such as albumin, collagen, gelatin and whey. These materials are composed of high molecular weight molecules existing in long chains which become intermeshed in random distribution to form a web or interlaced network or crossed fibers which provide structural fidelity to the compressed food product and enable it to return to its original geometry and size.

Also essential in the food products is a plasticizer such as a polyhydric alcohol, such as glycerol, propylene glycol, mannitol or sorbitol, which serve as a plasticizer for the protein molecules and enable them to be compressed and yet return to their original symmetry and form on admixture with water. The polyhydric alcohol must, of course, be water soluble and edible. Other plasticizers can be used in baked products which are insoluble in water, such as edible mono- and diglycerides, for example, glycerol monostearate and distearate.

In the practice of the process, a natural food product such as meat, fish, fruit, vegetable and the like, is shredded and dispersed within the framework of the protein and polyhydric alcohol components. The meat is usually shredded and homogenized with a water suspension of the protein containing the polyhydric alcohol. The resulting product is of a doughy consistency which can be aerated or whipped with mechanical equipment to introduce voids containing air, or it can be leavened with a leavening agent such as baking powder. The product is then shaped into desired form and cooked or baked.

Thereafter, the cooked or baked product is air-dried to a low moisture content. The moisture content can be thereafter adjusted to less than 5% by weight (by humidification if necessary) and then the product is compressed at high pressures (1,000 to 4,000 psi). The compression step results in a reduction of 10 to 50% of the original volume. The compressed product is then dried to a moisture content not greater than 3%, preferably not greater than 2%. Such food products can be stored at ordinary temperature if protected from moisture, or under refrigeration conditions or in the frozen state for prolonged periods of time.

Baked goods can also be comminuted and dispersed with water-insoluble plasticizers such as edible mono- and triglycerides to form doughy materials which can thereafter be dried and compressed. Included are bread, pastry, puddings, cakes and doughnuts.

Example 1: Compressed beef was produced from the following ingredients.

Component	Percent by Weight
Gluten	15
Soy protein (Promine D)	11
Propylene glycol	7
Beef steak	10
Water	57

The beef steak was fried and then ground in a meat grinder. It was dispersed in water in a Waring blender. Then, with continuous mixing, the propylene glycol was added, followed by the gluten (wetted with alcohol) and the soy protein. The entire dough mix was kneaded for approximately 5 minutes and shaped into the desired form, and left to relax for approximately 15 minutes. The specimens were then baked in a preheated oven at 350°F for 2 hours.

They were removed and air-dried with moisture level of not more than 2% by weight. To facilitate drying in some cases, the outer cooked layer of the baked product was removed. The dried specimens were then rehumidified in a steam environment to a moisture level of 4.5%. They were then compressed under a pressure of 2,000 psi, resulting in a three to one size reduction. The products were then air-dried at room temperature to a moisture content not greater than 2%.

The food products thus prepared could be stored at room temperature for an indefinite period. They were reconstituted by immersion in cold water and in hot water, each specimen expanding to its original size and form. The time required for full expansion varied from several seconds in hot water up to five minutes in cold water. Alternatively, the food products were expanded in a steam atmosphere and in hot oil. In each case, the product expanded to its original size and geometry.

Example 2: A food product similar to that described in Example 1 was made but edible cellulose fiber was substituted for the meat. The fiber contributes both to the mouth-feel texture of the product and to the maintenance of its structural integrity during compression. The recipe is as follows.

Component	Percent by Weight
Cellulose fiber	5
Glycerin	10
Nonfat dry milk solids	15
Gluten	40
Promine D	30

One part of this mixture was combined with two parts of water in preparing the product.

Example 3: A product prepared similar to Example 1 but not containing any meat or other fibrous material was made according to the following recipe.

Component	Percent by Weight
Gluten	50
Promine D	32
Nonfat dry milk solids	10.5
Propylene glycol	7.5

One part of this mixture was added to two parts of water in preparing the product.

BAR WITH OIL OR FAT BINDER

J.R. Durst; U.S. Patent 3,431,112; March 4, 1969; assigned to The Pillsbury Company describes a compact, highly nutritious food unit adapted to be eaten without further preparations or dispersed in water to form a soup, in which edible particles are held together with a binder. The binder is comprised of an edible oil or a normally solid fat, a film former and water.

The product comprises a binder consisting of a film former as a continuous phase or encapsulating material and a normally solid fat or oil. In one form of the process, edible food particles are distributed through the binder. The edible food particles may be either in flake, shredded, fibrous or powdered form and they provide the primary flavor and texture of the food bar.

Examples of edible food particles which can be used are: corn flakes, wheat flakes, rice, oats, graham cracker pieces, Rice Krispies, potato flakes, dried meat, vegetables, chocolate flakes or particles, cheese particles, ground peanuts, meat particles, raisins, dried fruit particles, fish, pregelatinized tapioca starch, and seasonings such as onion particles, pepper, salt, celery and monosodium glutamate. Other edible particles may be used as long as the particles are in flake, shredded, fibrous or particulate form.

The dispersion or binder is made up of an edible oil or a normally solid fat which is melted during formulation and a film former. Water is used during formulation but much or all of it is ordinarily removed to form the finished product. The edible oil may consist of any edible vegetable or animal oil or mixtures therein or normally solid fat and includes cottonseed oil, corn oil, lard, peanut oil, soy oil, safflower oil, butter or margarine.

The film former may consist of any edible substance that will form a film around the edible oil using any known process, as for example, vigorous mixing in an aqueous suspension, coacervation, spray drying a fat suspension in a film former and water solution, or by coating fat particles that have been chilled to a hard and nontacky condition. Other methods will be apparent to those skilled in the art.

Film formers include nonfat milk solids, sodium caseinate, soy protein, egg albumen, egg yolk, wheat germ, gelatin, pea flour, bean flour, corn germ, agar-agar, whey, gelatinized starch, fish protein, bran protein, gum arabic and other hydrophilic colloids, such as carboxymethylcellulose. Minor amounts of modifiers can be added to the film former if desired. Among such modifiers are salts, polysaccharides such as sucrose or lactose, and polyhydric alcohols such as glycerin.

Water is used to plasticize the film former. With the film former in a plastic state, vigorous mixing of the oil and film former results in the formation of an oil and film former dispersion. The dispersion consists of fat globules encapsulated in the film former. The oil is the discontinuous phase and the film former is the continuous phase. The dispersion serves as a binder in the formation of the food bar from edible particles.

Because the fat is encapsulated, there is no release of fat when the food bar is placed in water or milk to make a soup. Thus, the oil, film former, water mixture must be mixed until encapsulation is complete. A convenient test for determining when the constituents have been sufficiently mixed is a follows: remove one drop of the dispersion or binder and place it in 250 ml of water at 140°F. If fat is released, mixing is not complete and should be continued until the test procedure may be followed without the release of fat. The release of fat in the water system is a visual determination that can be readily made.

The proportions of the various constituents (edible particles, binder and water) may vary considerably. The proportion of the constituents (oil and film former) making up the binder may also vary. In general, however, the upper limit of edible particles is about 80% by weight of the food bar. The preferred upper limit is about 75%. The lower limit of edible particles in the food bar is about 10% when the edible particles comprise a nutritive food or a filler and about 0.25 to 0.5% of the weight of the bar when the edible particles comprise a flavoring or flavor imparting concentrate, but it is preferred that the bar contain from 30 to 75% by weight of edible particles.

Most preferred bar compositions have a density greater than 0.5 g/cc. Water is preferably used in the amount of about 5% by weight of the bar and preferably from 2 to 6%. Binder (oil and film former) should be provided in a quantity that comprises about 20 to 99.75% by weight of the bar.

The upper limit of fat in the binder is about 80% of the weight of the binder and the lower limit is about 10%. The dry weight of the film former exclusive of fillers as extenders can comprise from 5 to 90% of the dry weight of the binder, and preferably between 15 and 35% of the binder. Water should be present in the amount of from about 3% if the binder is dry to 50% if the binder is hydrated. Generally, it is preferred that the finished bar contain less than 10% moisture and the optimum moisture content is about 5% by weight based on the total weight of the food bar. However, up to 20% can be used if desired, but when such large amounts of water are used, enzymatic or bacterial degradation can be a problem in long term storage.

In one preferred manufacturing process an aqueous dispersion is made by adding the film former, fat and sucrose to a minimum amount of water (the least amount required to produce a suspension) which is ordinarily about 15 parts of water per 100 parts of solids. Mixing can be carried out in any conventional mixer such as a Hamilton Beach or Hobart mixer until a thick gel is formed. When food particles are used in the formulation, about 20% of the hydrated binder is mixed with about 80% of the solid food particles on a weight basis, and the resulting mixture is either formed or subjected to substantial pressure if greater hardness is desired into the shape of a food bar. The product is then ready for packaging but, if desired, it can be dried by placing it in a oven prior to packaging.

In another preferred process, the preparation of the binder is begun in the same manner described in the preceeding paragraph except that 50 parts of water are used rather than 15 parts. More or less water can be used to properly adjust the viscosity of the resulting dispersion until it can be pumped to a spray dryer of a suitable construction to produce a white, free-flowing powder. If food particles are used in the composition, the powder is dry blended with the food particles at this point.

The binder, with or without food particles, is then mixed with the minimum amount of water required to cause the surfaces thereof to become adherent, i.e., tacky. The particles are then placed in a press and subjected to pressure to form a food bar. Since the moisture level of the finished bar is relatively low, no further drying is required in most instances. Moreover, the hardness of the finished product can be precisely controlled. The product formed in accordance with this second manufacturing procedure also has the advantage of rehydrating more readily and more quickly. As a result, it can be more easily chewed and more easily mixed with water if it is to be used as a rehydrated food such as soup, pudding or a beverage as the case may be.

When the bar is to be formed, the binder and edible food particles are mixed and placed in a mold cavity of suitable known construction. The molding pressure can vary from about 25 to 1,000 psi and preferably between 250 to 750 psi. The molding time can vary from about 0.25 second to any desired time period. For most purposes, from 1 to 10 seconds is preferred.

The process may be appropriately illustrated by the following examples in which all amounts are set forth as percentages by weight.

Example 1: The following binder formulation was made.

Nonfat milk solids	14.8%
Lard flakes	14.8%
Sucrose	18.7%
Water	51.7%

The lard flakes were heated in a steam jacketed kettle to 160°F to completely melt them. The nonfat milk solids and sucrose were added and mixed with the melted lard flakes. About half (53.6%) of the water was added with rapid agitation. The mixture was then pumped through an impeller mixer known as an Oakes mixer and agitated for 35 minutes. The remainder of the water was added to reduce the viscosity for spray-drying. The material was pumped under pressure of 1,000 to 1,200 psi through a spray nozzle of a horizontal Blaw-Knox spray dryer. The inlet air temperature was 230° to 240°F and the outlet air temperature was 170° to 175°F. The resulting product was a stable, free-flowing white powder.

The spray-dried dispersion was used to form a food bar consisting of: edible particles, 52%; binder, 45%; water, 3% according to the following procedure.

Food bars were made in which corn flakes, wheat flakes, Rice Krispies, and graham crackers comprised the edible particles. The edible particles and binder were mixed at high speed, using a Hobart mixer for a period of three to four minutes, until the maximum dimension of the edible particles was 1/16" to 1/8". The Hobart mixer was set at No. 2 speed and the water was slowly added while mixing continued. 40 gram units of the resulting free-flowing mixture were placed into 2" x 4" dies and pressed into a bar under 125 psi pressure. Upon release from the die, the bars were satisfactory, but after drying for 20 minutes in an air circulating oven at 50°C, their cohesive strength was improved.

Example 2: A binder formulation was made as follows.

Nonfat milk solids	25.8%
Cottonseed oil	25.8%
Sucrose	25.8%
Glycerin	6.5%
Water	16.1%

The stable dispersion was formed by placing the cottonseed oil in a Waring Blendor, adding the nonfat milk solids and sucrose, and mixing. The glycerin was dissolved in the water and the solution was added to the material in the mixer. A stable dispersion was formed with continued mixing at high speed for one minute.

A corn flake bar was made with the above dispersion as follows: 34.8 parts of the above dispersion was mixed with 65.2 parts of corn flakes in a Hobart mixer. The materials were mixed at No. 3 speed for two minutes. 30 gram units were placed into a 2" x 4" mold and subjected to 250 psi pressure. After removal from the mold, the bars were palatable and could be eaten "as is" or broken up and added to water. The bar contained four calories per gram and 7.9% water.

A second bar was made with the same dispersion as a binder in the ratio of one part binder to one part corn flakes. The same procedure was followed. The resulting food bar contained 4.5% moisture and 4.4 calories per gram following drying in a hot air dryer.

Example 3: A binder was made using a higher percentage of fat and using sodium caseinate as the film former. The formulation was as follows.

Lard flakes	35.7%
Distilled water	28.6%
Sucrose	22.5%
Sodium caseinate	8.6%
Glycerin	4.6%

The above binder was used to make a hash bar having the following formulation.

Ingredient	%
Binder	38.6%
Potato flakes	38.4%
Dried gravy mix	13.8%
Dried beef (oil immersion dried)	8.2%
Onion flakes	1.0%

The above ingredients were mixed together in a Hobart mixer at high speed for about two minutes. Food bars were made following the procedure stated in Example 2. The bars were dried to 4.5% moisture and contained 4.6 calories per gram and 18.2% fat. The bar was palatable when eaten "as is" or when added to one cup of hot water.

Example 4: A binder was made with the following formulation.

Ingredient	%
Sucrose	6.92%
Durkex oil	21.43%
Nonfat milk solids	21.57%
Water	50.08%

The constituents were mixed and spray-dried according to the procedure stated in Example 1, except that the dispersion was homogenized at 4,000 psi before spray-drying. The stable, white, free-flowing powders resulting were used to make a hash bar, a potato soup with beef bar, and a split pea bar.

A hash bar was made according to the following formulation.

Ingredient	%
Potato flakes	43.0%
Binder	35.5%
Dried beef	8.2%
Dried gravy mix	7.2%
Water	5.0%
Dried onion flakes	0.5%
Sodium chloride	0.5%
Seasoning	0.1%

The dried beef was reduced in size by mixing in a Hobart mixer and then the potato flakes were added, followed by the seasoning and the binder. Mixing was continued at No. 2 speed for one minute. The water was then added and mixing continued until a uniform mixture was obtained. The product was made into 2" x 4" x ¼" bars weighing approximately 40 grams. The pressure was 750 psi and the dwell time was five seconds. The bars were dried for 20 minutes in an air circulating oven at 50°C. The dried bars were packaged into Mylar-vinyl lined aluminum foil pouches. The bars contained 7.8% moisture, 4.4 calories per gram and 16.3% fat.

The binder was used to make a potato soup with beef bar, according to the following formulation.

Potato flakes	45.2%
Binder	39.4%
Dried beef	8.2%
Water	5.0%
Sodium chloride	1.0%
Dried onion powder	0.5%
Dried celery	0.6%
Seasoning	0.1%

The bar was made according to the procedure stated immediately above, in connection with the hash bar. The bars contained 7.4% moisture, 4.4 calories per gram and 18.0% fat.

A split pea bar was made according to the following formulation.

Binder	46.8%
Split pea powder	30.8%
Potato granules	7.8%
Potato flakes	7.8%
Water	4.0%
Sodium chloride	1.0%
Smoked yeast flour	1.0%
Onion powder	0.5%
Monosodium glutamate	0.2%
Black pepper	0.1%

The procedure used for making the bar was the same as that stated above in connection with the hash bar, except the dwell time was 0.3 second instead of 5 seconds. The bars contained 7.1% moisture, 4.5 calories per gram and 20.9% fat.

Example 5: A binder was made with the following formulation.

Lactose	15.14%
Durkex oil	5.71%
Sodium caseinate	7.61%
Water	71.54%

The Durkex oil was heated to 150°F and placed into a Waring Blendor. The sodium caseinate was added to the oil and 60% of the water was added at 140°F. Mixing continued for one minute at high speed. The lactose, dissolved in the remainder of the water, was then added. The material was sprayed with a Bowen spray dryer with a chamber temperature of 180°F. A white, free-flowing powder resulted.

A food bar was made according to the following formulation.

Freeze-dried scrambled eggs	70%
Binder	25%
Water	5%

The constituents were thoroughly mixed and pressed into a 2" x 4" x ½" mold under 250 psi pressure.

Example 6: The spray-dried binder of Example 5 was also used in the following formulation.

Fried bacon pieces	70%
Binder	25%
Water	5%

Acceptable bars were made according to the procedure of Example 5.

Example 7: Dried fish bars were made with the binder of Example 5, according to the following formulation.

Dried smoked herring	70%
Binder	25%
Water	5%

The herring was broken into small pieces in a Waring Blendor and then placed into a Hobart mixer. The water was added slowly and then the binder was added with continued mixing. After a uniform mixture was obtained, the product was prepared into bars that were very satisfactory.

Example 8: A binder was made according to the following formulation.

Sucrose	3.21%
Durkex oil	21.61%
Sodium caseinate	8.81%
Lactose	7.66%
Cornstarch	4.59%
Water	54.12%

The Durkex oil was heated to 150°F and placed into a Waring Blendor. The sucrose, starch and sodium caseinate were added and coated with Durkex oil. Sixty percent of the water at 150°F was added with high speed mixing and the mixing continued for 1.5 minutes. The lactose was dissolved in the balance of the water and added. Mixing continued for one minute at high speed. The dispersion was spray-dried (Bowen) with a chamber temperature of 170°F. A white, free-flowing powder resulted.

Food bars were made according to the following formulation.

Roasted peanut pieces	70%
Binder	25%
Water	5%

The peanuts were chopped in a Waring Blendor and screened. Only particles smaller than No. 7 mesh (U.S. Sieve Series) were used. The particles were placed in a Hobart mixer and the water added slowly while mixing at No. 2 speed. The binder was added and mixing continued for 30 seconds. Bars were made and dried 20 minutes in an air circulating oven at 50°C. The resulting bars were acceptable.

Example 9: The binder of Example 8 was used to make food bars according to the following formulation.

Freeze-dried chicken	62.50%
Binder	20.83%
Glycerin	2.67%
Water	14.00%

The chicken was weighed into a 12 quart bowl and mixed at No. 2 speed in a Hobart mixer for one minute. The binder was added and mixing continued for 30 seconds. The glycerin-water solution was added and mixing continued for one minute. 120 bars at 2" x 4" x ½" were made using 750 psi pressure and a dwell time of 10 seconds. The bars were then removed from the press and dried for 3.5 hours at 120°F. The resulting bars were of a considerable hardness and did not break when subjected to the 6 foot drop test until the fourth drop.

Example 10: The binder of Example 8 was used to make a food bar according to the following formulation.

Dried peaches	71%
Binder	25%
Water	5%

The dried peach halves were chopped, using a Waring Blendor, and water was added to the chopped peach halves while mixing at No. 2 speed. The binder was added and mixing continued until a uniform mixture was obtained. 120 bars at 2" x 4" x ½" were made by utilizing a pressure of 312.5 psi and a dwell time of 10 seconds. The bars were dried for 30 minutes at 120°F. The resulting bars were of an acceptable hardness and did not break when subjected to the 6 foot drop test until the eleventh drop.

DEHYDRATION OF FRUITS AND VEGETABLES 1974

by M. Torrey

Food Technology Review No. 13

Without appropriate dehydration techniques most convenience foods, as we know and eat them today, would not exist. Dehydration is also an integral part of agglomeration processes, so essential in controlling bulk density, dispersibility and solubility of powdered food products.

Dehydration, i.e. removing the water content of foods by as much as 99%, makes it impossible for most microorganisms and enzymes to digest the food and to propagate on it. Dehydration also reduces other deteriorative changes, such as chemical oxidation, usually evidenced by undesirable changes in color and taste. Refrigeration of dehydrated foods is unnecessary in most instances.

Another important consideration is the reduction in shipping weight and volume, and the corresponding savings in cost of transportation. Water is available almost everywhere, and rehydration of most foods is accomplished easily, provided proper procedures are followed.

The present state of the art, as it applies to fruits and vegetables, is described in this book by means of 206 processes gleaned from recent U.S. patents.

A partial and condensed table of contents follows here. Numbers in parentheses indicate the number of processes per topic. Chapter headings are given, followed by examples of important subtitles.

ISBN 0-8155-0527-2

287 pages